Rao / O'Brien

Design Data for Plastics Engineers

Natti Rao / Keith O'Brien

Design Data for Plastics Engineers

Hanser Publishers, Munich
Hanser/Gardner Publications, Inc., Cincinnati

The Authors:
Dr.-Ing. Natti Rao, Schieferkopf 6, 67434 Neustadt, Germany; *Dr. Keith T. O'Brien,* Vistakon, Ing., 4500 Salisbury Road, Jacksonville, FL 32216-0995, USA

Distributed in the USA and in Canada by
Hanser/Gardner Publications, Inc.
6915 Valley Avenue, Cincinnati, Ohio 45244-3029, USA
Fax: (513) 527-8950
Phone: (513) 527-8977 or 1-800-950-8977
Internet: http://www.hansergardner.com

Distributed in all other countries by
Carl Hanser Verlag
Postfach 86 04 20, 81631 München, Germany
Fax: +49 (89) 98 12 64

Library of Congress Cataloging-in-Publication Data
Rao, Natti S.
Design data for plastics engineers / Natti Rao, Keith O'Brien.
p. cm.
Includes bibliographical references and index.
ISBN 1-56990-264-X (softcover)
1. Plastics. I. O'Brien, Keith, T. II. Title.
TP1120.R325 1998
668.4--dc21 98-37253

Die Deutsche Bibliothek - CIP-Einheitsaufnahme
Rao, Natti S.:
Design data for plastics engineers / Natti Rao/Keith O'Brien. -
Munich : Hanser ; Cincinnati : Hanser/Gardner, 1998
ISBN 3-446-21010-5

Camera-ready copy prepared by the authors.
Printed and bound in Germany by Druckhaus "Thomas Müntzer", Bad Langensalza

Preface

Mechanical, thermal and rheological properties of polymers form the basic group of property values required for designing polymer machinery. In addition, knowledge of the properties of the resin such as stock temperature of the melt is necessary for optimizing the process. Furthermore, while designing a plastic part, performance properties of the resin depending on the application are to be considered, examples of which are flammability, weather resistance and optical properties, to quote a few. Hence, a variety of property values is needed to accomplish machine design, part design and process optimization.

Comprehensive as well as specific overviews of polymer property data exist in the literature. Also the data banks of resin manufacturers offer quick information on thermophysical properties of polymers. But there are few books dealing with both resin and machine design data based on modern findings.

The intent of this book is first of all to create an easy to use quick reference work covering basic design data on resin, machine, part and process, and secondly, to show as to how this data can be applied to solve practical problems. With this aim in mind numerous examples are given to illustrate the use of this data. The calculations involved in these examples can be easily handled with the help of handheld calculators.

Chapters 1 to 5 deal with the description of physical properties – mechanical, thermal, rheological, electrical and optical – of polymers and principles of their measurement. In Chapter 6 the effect of external influences on the performance of polymers is treated. General property data for different materials such as liquid crystal polymers, structural foams, thermosetting resins and reinforced plastics are given at the end of this chapter.

In Chapter 7 the processing properties and machine related data are presented for continuous extrusion processes namely blown film, pipe and flat film extrusion. Resin and machine parameters for thermoforming and compounding have also been included in this chapter.

Chapter 8 deals with blow molding and the influence of resin and machine variables on different kinds of blow molding processes. Finally, Chapter 9 covers resin-dependent and machine related parameters concerning the injection molding process.

Machine element design covered in Chapters 7 to 9 includes screw design for extruders and injection molding machines, die design for extruders, mold design for molding and forming operations and downstream equipment for extrusion. Wherever appropriate, the properties and machine related parameters are described by mathematical formulas which are, as already mentioned, illustrated by worked-out examples. The solution procedure used in these examples describes the application of polymer data to solve practical problems. On the basis of this approach the importance of polymer data in dealing with design and process optimization is explained.

This book is intended for beginners as well as for practicing engineers, students and teachers in the field of plastics technology and also for scientists from other fields who deal with polymer engineering.

We wish to express our sincere thanks to Professor Stephen Orroth of the University of Massachusetts at Lowell, U.S.A., for his valuable suggestions, criticism and review of the manuscript. He also suggested to write a book on the lines of Glanvill's Plastics Engineer's Data Book. Our work is based on the findings of modern plastics technology, and can be considered as an extension to Glanvill's book.

The authors also wish to express particular thanks to Dr. Günter Schumacher of University of Karlsruhe, Germany, for his constructive comments and his great help in preparing the manuscript.

Neustadt, Germany
Jacksonville, U.S.A.

Natti S. Rao, Ph.D.
Keith T. O'Brien, Ph.D.

Contents

1 Mechanical Properties of Solid Polymers

The properties of polymers are required firstly to select a material which enables desired performance of the plastics component under conditions of its application. Furthermore they are also essential in design work to dimension a part from a stress analysis or to predict the performance of a part under different stress situations involved. Knowledge of polymer properties is, as already mentioned in the preface, a prerequisite for designing and optimizing polymer processing machinery.

In addition to the physical properties there are certain properties known as performance or engineering properties which correlate with the performance of the polymer under varied type of loading and environmental influences such as impact, fatigue, high and low temperature behavior and chemical resistance. The following sections deal with the physical as well as important performance properties of polymers.

1.1 Ideal Solids

Ideal elastic solids deform according to Hookean law which states that the stress is directly proportional to strain. The behavior of a polymer subjected to shear or tension can be described by comparing its reaction to an external force with that of an elastic solid under load. To characterize ideal solids, it is necessary to define certain quantities as follows [3]:

1.2 Tensile Properties

The axial force F_n in Fig. 1.1 causes an elongation Δl of the sample of diameter d_0 and length l_0 fixed at one end. Following equations apply for this case:
Engineering strain:

$$\varepsilon' = \frac{\Delta l}{l_0} \qquad (1.2.1)$$

Hencky strain:

$$\varepsilon = \ln\left(\frac{l}{l_0}\right) \tag{1.2.2}$$

Tensile stress:

$$\sigma_Z = \frac{F_n}{A_0} \tag{1.2.3}$$

Reference area:

$$A_0 = \frac{\pi d_0^2}{4} \tag{1.2.4}$$

Poisson's ratio:

$$\mu = -\frac{\Delta d / d_0}{\Delta l / l_0} \tag{1.2.5}$$

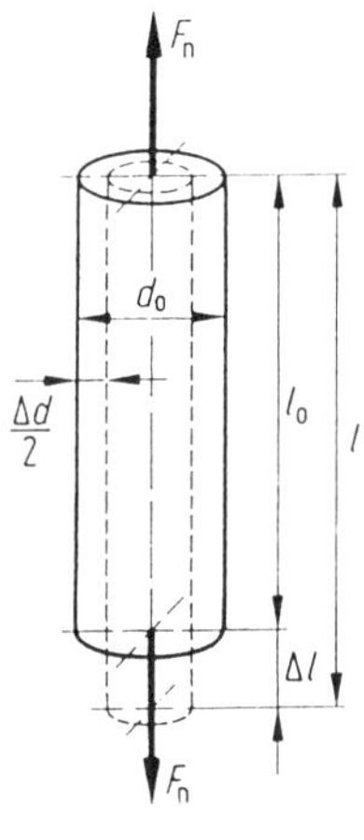

Fig. 1.1: Deformation of a Hookean solid by a tensile stress [10]

1.2.1 Stress-Strain Behavior

As shown in Fig. 1.2 most metals exhibit a linear stress-strain relationship (curve 1), where as polymers being viscoelastic show a non-linear behavior (curve 2). When the stress is directly proportional to strain the material is said to obey Hooke's law. The slope of the straight line portion of curve 1 is equal to the modulus of elasticity. The maximum stress point on the curve, up to which stress and strain remain proportional is called the proportional limit (point P in Fig. 1.2) [1].

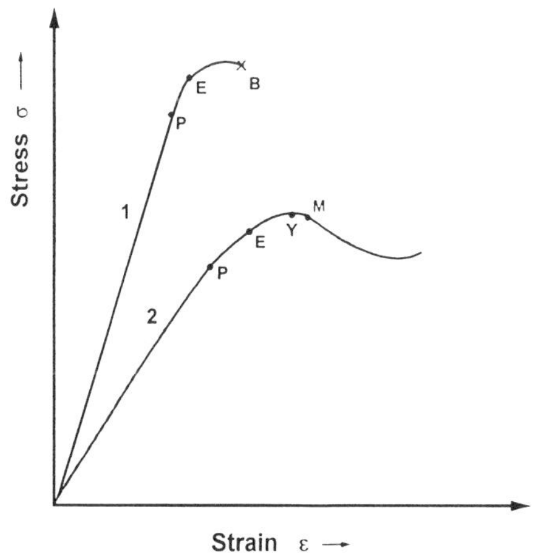

Fig. 1.2: Typical stress-strain curves of metals and polymers

Most materials return to their original size and shape, even if the external load exceeds the proportional limit. The elastic limit represented by the point E in Fig. 1.2 is the maximum load which may be applied without leaving any permanent deformation of the material. If the material is loaded beyond its elastic limit, it does not return to its original size and shape, and is said to have been permanently deformed. On continued loading a point is reached at which the material starts yielding. This point (point Y in Fig. 1.2) is known as the yield point, where an increase in strain occurs without an increase in stress. It should however be noted that some materials may not exhibit a yield point. The point B in Fig. 1.2 represents break of the material.

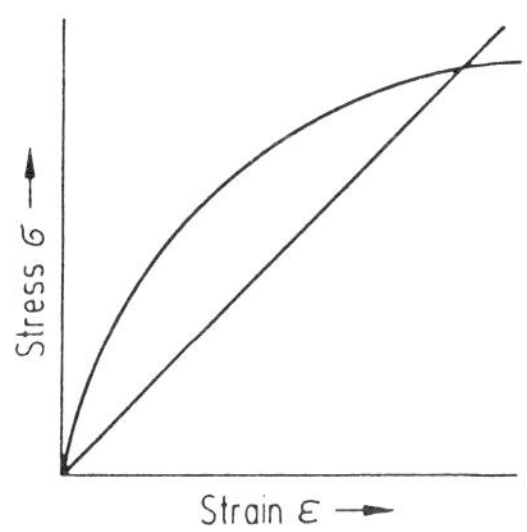

Fig. 1.3: Secant modulus [3]

Owing to their non-linear nature it is difficult to locate the straight line portion of the stress-strain curve for polymeric materials (Fig. 1.3). The secant modulus represents the ratio of stress to strain at any point (S in Fig. 1.3) on the stress-strain diagram and is equal to the slope of the line OS. It is an approximation to a linear response over a narrow, but pre-specified and standard level of strain [2], which is usually 0.2%.

The initial modulus is a straight line drawn tangent to the initial region of the stress-strain curve to obtain a fictive modulus as shown in Fig. 1.4. As this is ambiguous, the resin manufacturers provide a modulus with the stress, for example, $\sigma_{0.5}$ corresponding to a strain 0.5 % to characterize the material behavior on a practical basis (Fig. 1.4) [3]. This stress $\sigma_{0.}$ is also known as the proof stress.

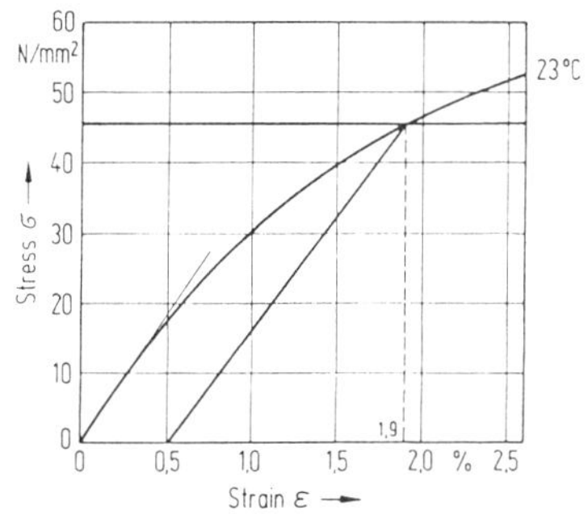

Fig. 1.4: Stress at a strain of 0.5% [3]

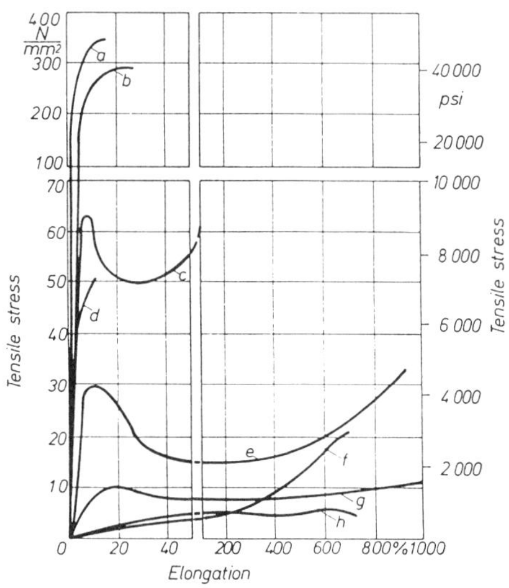

Fig. 1.5: Tensile stress diagram of a number of materials at 23°C [4]: a: steel, b: copper, c: polycarbonate, d: PMMA, e: PE-HD, f: rubber, g: PE-LD, h: PVC-P

Stress-strain diagrams are given in Fig. 1.5 for a number of materials [4]. It can be seen from Fig. 1.5 that the advantage of metals lies in their high strength, where as that of plastics lies in their high elongation at break.

1.2.2 Tensile Modulus

According to Eq. (1.2.1) and Eq. (1.2.3) one obtains for the modulus of elasticity E which is known as Young's Modulus

$$E = \sigma_z / \varepsilon \qquad (1.2.6)$$

The modulus of elasticity in a tension test is given in Table 1.1 for different polymers [7].

Table 1.1: Guide Values of Modulus of Elasticity of Some Plastics [4]

Material	Modulus of elasticity $N\,mm^{-2}$
PE-LD	200/500
PE-HD	700/1400
PP	1100/1300
PVC-U	1000/3500
PS	3200/3500
ABS	1900/2700
PC	2100/2500
POM	2800/3500
PA6	1200/1400
PA66	1500/2000
PMMA	2700/3200
PET	2600/3100
PBT	1600/2000
PSU	2600/2750
CA	1800/2200
CAB	1300/1600
Phenol-Formaldehyde-Resins	5600/12000
Urea-Formaldehyde-Resins	7000/10500
Melamin-Formaldehyde-Resins	4900/9100
Unsaturated Polyester Resins	14000/20000

The 3.5 % flexural stresses of thermoplastics obtained on a 3-point bending fixture (Fig. 1.6) [6] lie in the range 100 to 150 N mm^{-2} and those of thermosets from around 60 to 150 N mm^{-2} [7].

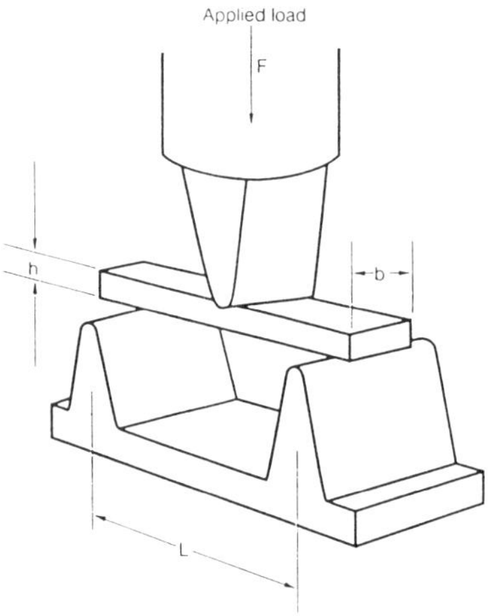

Fig. 1.6: Three point bending fixture [6]

1.2.3 Effect of Temperature on Tensile Strength

The tensile strength is obtained by dividing the maximum load (point M in Fig. 1.2) the specimen under test will withstand by the original area of cross-section of the specimen. Fig. 1.7 shows the temperature dependence of the tensile strength of a number of plastics [4].

1.3 Shear Properties

Figure 1.8 shows the influence of a shear force F_t acting on the area A of a rectangular sample and causing the displacement ΔU. The valid expressions are defined by:

Shear strain:

$$\gamma = \frac{\Delta U}{l} \tag{1.3.1}$$

Shear stress

$$\tau = \frac{F_t}{A} \tag{1.3.2}$$

1.3.1 Shear Modulus

The ratio of shear stress to shear strain represents the shear modulus G. From the equations above results:

$$G = \tau / \gamma \tag{1.3.3}$$

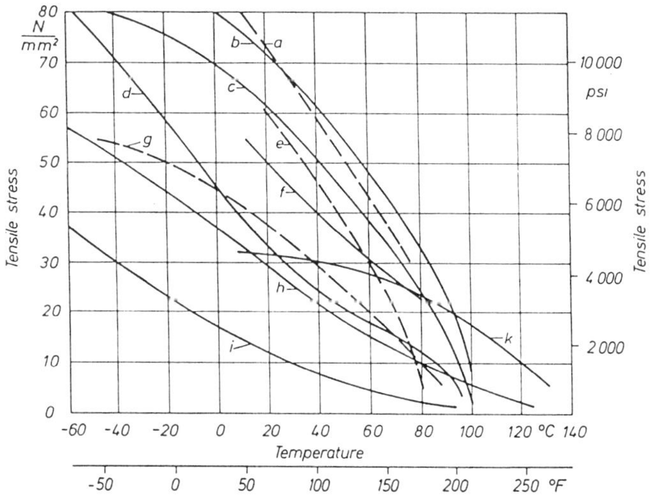

Fig. 1.7: Temperature dependence of the tensile stress of some thermoplastics under uniaxial loading [4]; a: PMMA, b: SAN, c: PS, d: SB, e: PVC-U f: ABS, g: CA, h: PE-HD, i: PE-LD, k: PE-LD-V

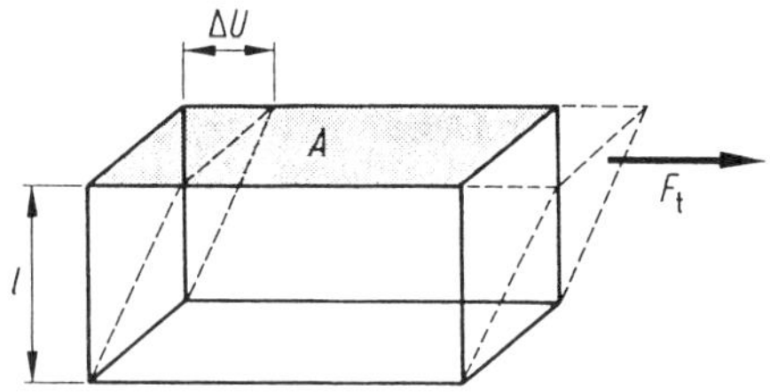

Fig. 1.8: Deformation of a Hookean solid by shearing stress [10]

1.3.2 Effect of Temperature on Shear Modulus

The viscoelastic properties of polymers over a wide range of temperatures can be better characterized by the complex shear modulus G^* which is measured in a torsion pendulum test by subjecting the specimen to an oscillatory deformation Fig. 1.9 [2]. The complex shear modulus G^* is given by the expression [2]

$$|G^*| = \left[(G')^2 + (G'')^2\right]^{0.5} \tag{1.3.4}$$

The storage modulus G' in Eq. (1.3.4) represents the elastic behavior associated with energy storage and is a function of shear amplitude, strain amplitude and the phase angle

δ between stress and strain. The loss modulus G'' which is a component of the complex modulus depicts the viscous behavior of the material and arises due to viscous dissipation.

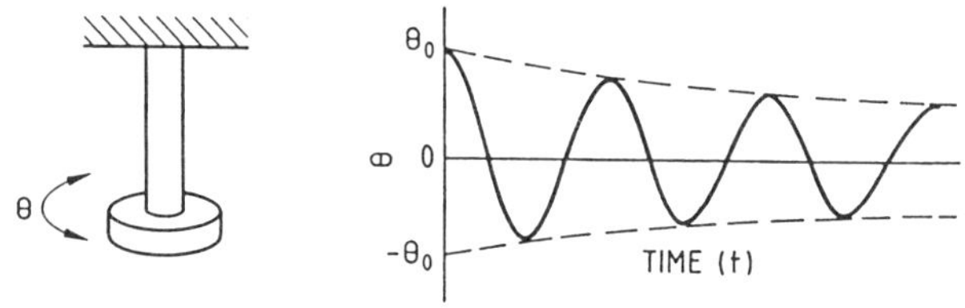

Fig. 1.9: Torsion pendulum; Loading mode and sinusoidal angular displacement t [2]

The tangent of the phase angle δ is often used to characterize viscoelastic behavior and is known as loss factor. The loss factor d can be obtained from

$$d = \frac{G''}{G'} = \tan\delta \tag{1.3.5}$$

The modulus-temperature relationship is represented schematically in Fig. 1.10 [2], from which the influence of the transition regions described by the glass transition temperature T_g and melting point T_m is evident.

This kind of data provides information on the molecular structure of the polymer. The storage modulus G' which is a component of the complex shear modulus G^* and the loss factor d are plotted as functions of temperature for high density polyethylene in Fig. 1.11 [4]. These data for various polymers are given in the book [4].

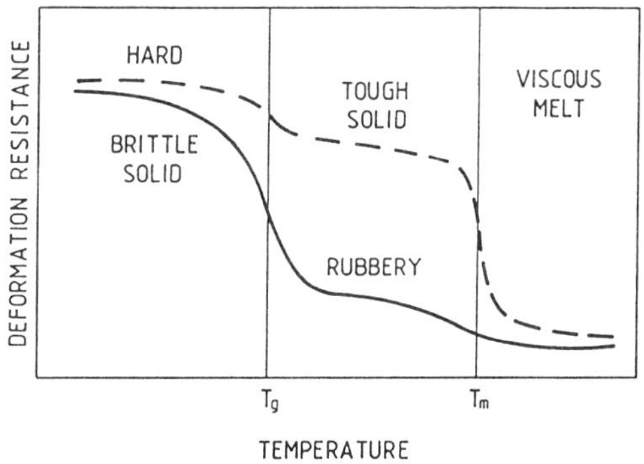

Fig. 1.10: Generalized relationship between deformation resistance and temperature for amorphous (solid line) and semi-crystalline (broken line) high polymers [2]

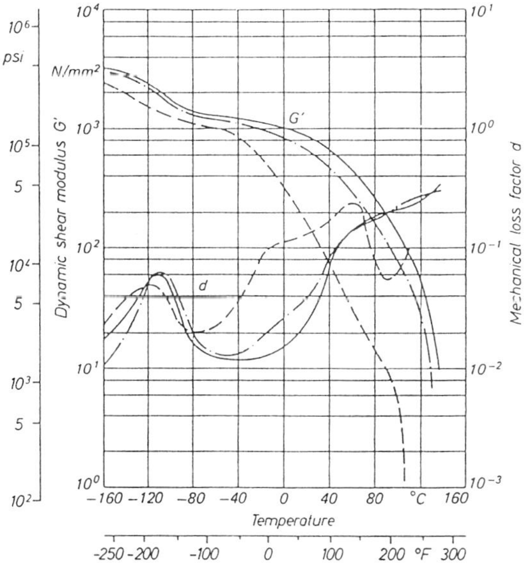

Fig. 1.11: Temperature dependence of the dynamic shear modulus, G' and the loss factor d, obtained in the torsion pendulum test DIN 53445 [4]: PE-HD (highly crystalline), PE-HD (crystalline), PE-LD (less crystalline)

1.4 Compressive Properties

The isotropic compression due to the pressure acting on all sides of the parallelepiped shown in Fig. 1.12 is given by the engineering compression ratio κ.

$$\kappa = \frac{\Delta V}{V_0} \tag{1.4.1}$$

where ΔV is the reduction of volume due to deformation of the body with the original volume V_0.

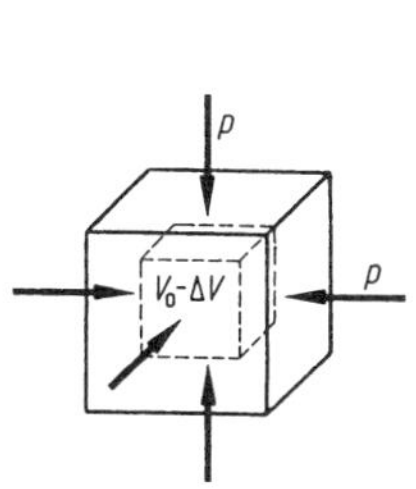

Fig. 1.12: Hookean solid under compression [44]

Table 1.2: Poisson ratio μ, density ρ, bulk modulus K and specific bulk modulus for some materials [10]

Material	Poisson ratio μ	Density ρ at 20°C g/cm^3	Bulk modulus K N/m^2	Specific bulk modulus K/ρ m^2/s^2
Mild steel	0.27	7.8	$1.66 \cdot 10^{11}$	$2.1 \cdot 10^{7}$
Aluminum	0.33	2.7	$7 \cdot 10^{10}$	$2.6 \cdot 10^{7}$
Copper	0.25	8.9	$1.34 \cdot 10^{11}$	$1.5 \cdot 10^{7}$
Quartz	0.07	2.65	$3.9 \cdot 10^{10}$	$1.47 \cdot 10^{7}$
Glass	0.23	2.5	$3.7 \cdot 10^{10}$	$1.49 \cdot 10^{7}$
Polystyrene	0.33	1.05	$3 \cdot 10^{9}$	$2.85 \cdot 10^{7}$
Polymethyl-methacrylate	0.33	1.17	$4.1 \cdot 10^{9}$	$3.5 \cdot 10^{7}$
Polyamide 66	0.33	1.08	$3.3 \cdot 10^{9}$	$2.3 \cdot 10^{7}$
Rubber	0.49	0.91	$0.033 \cdot 10^{9}$	$0.04 \cdot 10^{7}$
PE-LD	0.45	0.92	$0.7 \cdot 10^{9}$	$3.7 \cdot 10^{7}$
Water	0.5	1	$2 \cdot 10^{9}$	$2 \cdot 10^{6}$
Organic liquids	0.5	0.9	$1.33 \cdot 10^{9}$	$1.5 \cdot 10^{6}$

1.4.1 Bulk Modulus

The bulk modulus K is defined by

$$K = -p/\kappa \tag{1.4.2}$$

where κ is calculated from Eq. (1.4.1). The reduction of volume, for instance for PE-LD, when the pressure is increased by 100 bar follows from Table

$$\Delta V / V_0 = -100/(0.7 \cdot 10^4) = -1.43\% .$$

Furthermore, the relationship between E, G and K is expressed as [3]

$$E = 2\,G(1+\mu) = 3K(1-2\mu) . \tag{1.4.3}$$

This leads for an incompressible solid ($K \to \infty$, $\mu \to 0.5$) to

$$E = 3G . \tag{1.4.4}$$

Typical values of moduli and Poisson's ratios for some materials are given in Table 1.2 [10]. Although the moduli of polymers compared with those of metals are very low, at equal weights, i.e. ratio of modulus to density, polymers compare favorably.

1.5 Time Related Properties

1.5.1 Creep Modulus

In addition to stress and temperature, time is an important factor for characterizing the performance of plastics. Under the action of a constant load a polymeric material experiences a time dependent increase in strain called creep. Creep is therefore the result of increasing strain over time under constant load [6]. Creep behavior can be examined by subjecting the material to tensile, compressive or flexural stress and measuring the strain for a range of loads at a given temperature.

The creep modulus $E_c(t)$ in tension can be calculated from

$$E_c(t) = \tau / \varepsilon(t) \quad (1.5.1)$$

and is independent of stress only in the linear elastic region.

As shown in Fig. 1.13 the creep data can be represented by creep plots, from which the creep modulus according to Eq. (1.5.1) can be obtained. The time dependence of tensile creep modulus of some thermoplastics at 20°C is shown in Fig. 1.14 [5]. The 2 % elasticity limits of some thermoplastics under uniaxial stress are given in Fig. 1.15. Creep data measured under various conditions for different polymers are available in [4].

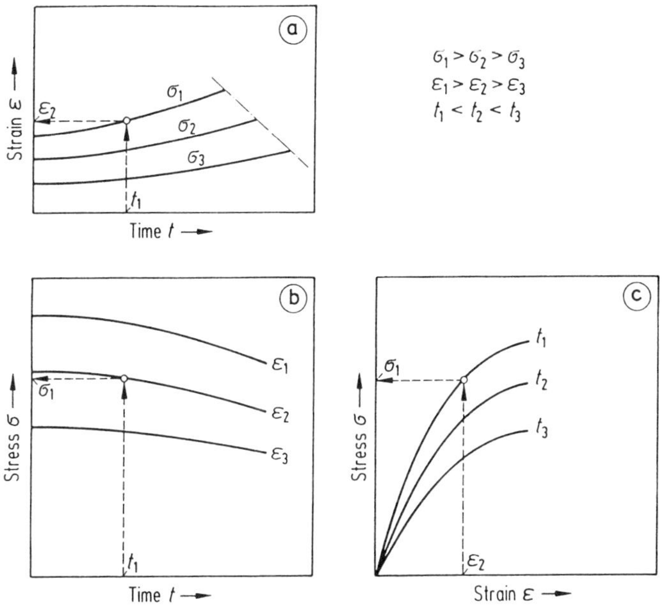

Fig. 1.13: Long-term stress-strain behavior [3]

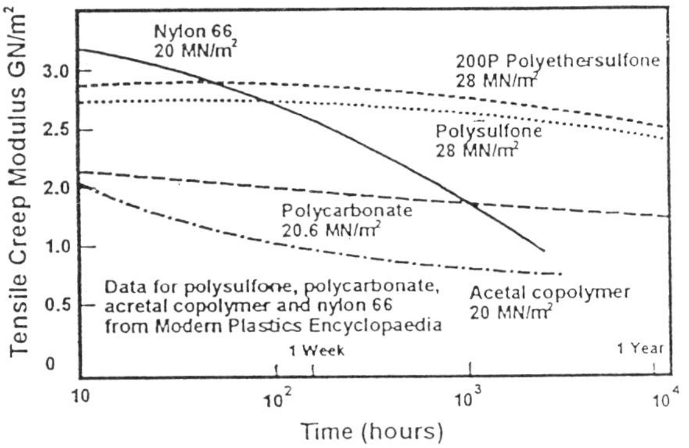

Fig. 1.14: Tensile creep modulus vs time for engineering thermoplastics [5]

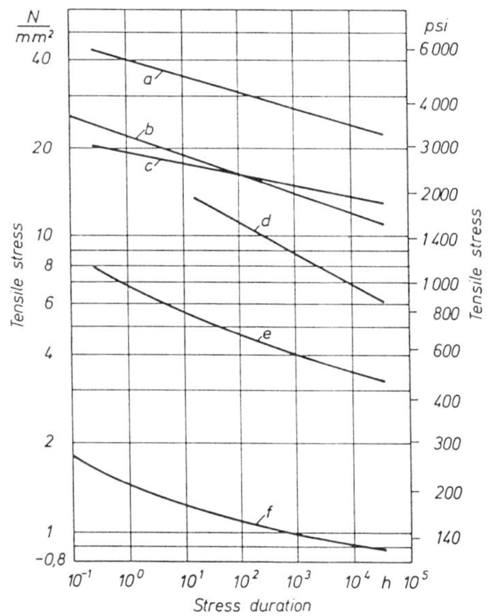

Fig. 1.15: 2% elasticity limits of some thermoplastics under uniaxial stress at 20°C [4]
a: SAN, b: ABS, c: SB, d: chlorinated polyether, e: PE-HD, f: PE-LD

1.5.2 Creep Rupture

Failure with creep can occur when a component exceeds an allowable deformation or when it fractures or ruptures [6]. Creep rupture curves are obtained in the same manner as creep, except that the magnitude of the stresses used is higher and the time is measured up to failure. Data on creep rupture for a number of polymers are presented in [4]. According to [4] the extrapolation of creep data should not exceed one unit of logarithmic time and a strain elongation limit of 20 % of the ultimate strength.

1.5.3 Relaxation Modulus

The relaxation behavior of a polymer is shown in Fig. 1.16 [3]. Relaxation is the stress reduction which occurs in a polymer when it is subjected to a constant strain. This data is of significance in the design of parts which are to undergo long-term deformation. The relaxation modulus of PE-HD-HMW as a function of stress duration is given in Fig. 1.17 [4]. Similar plots for different materials are to be found in [4].

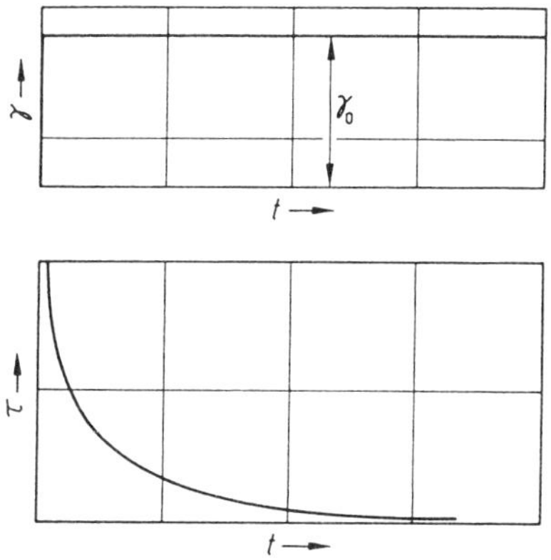

Fig. 1.16: Relaxation after step shear strain γ_0 [3]

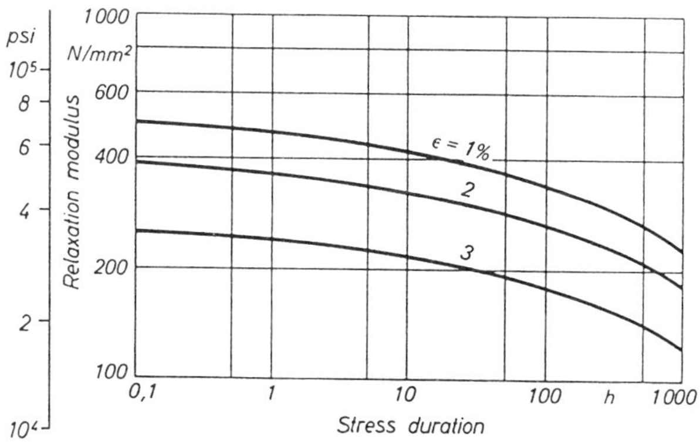

Fig. 1.17: Relaxation modulus of PE-HD-HMW as a function of stress duration at 23°C [4]

1.5.4 Fatigue Limit

Fatigue is a failure mechanism which results when the material is stressed repeatedly or when it is subjected to a cyclic load. Examples of fatigue situations are components subjected to vibration or repeated impacts. Cyclic loading can cause mechanical deterioration and fracture propagation resulting in ultimate failure of the material.

Fatigue is usually measured under conditions of bending where the specimen is subjected to constant deflection at constant frequency until failure occurs. The asymptotic value of stress shown in the schematic fatigue curve (S-N plot) in Figs. 1.18 and 1.19 [6] is known as the fatigue limit. At stresses or strains which are less than this value failure does not occur normally.

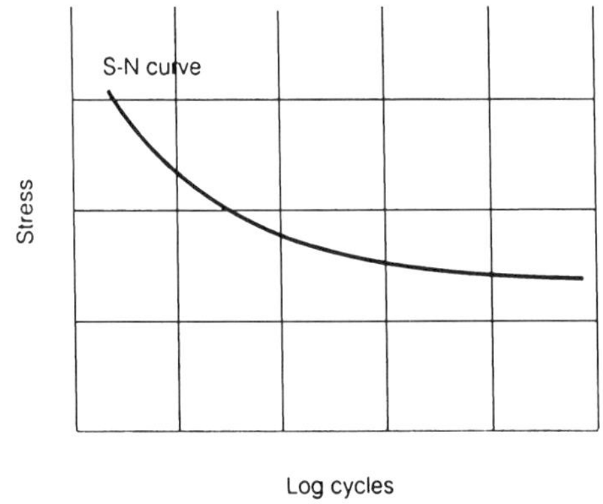

Fig. 1.18: Fatigue curve [6]

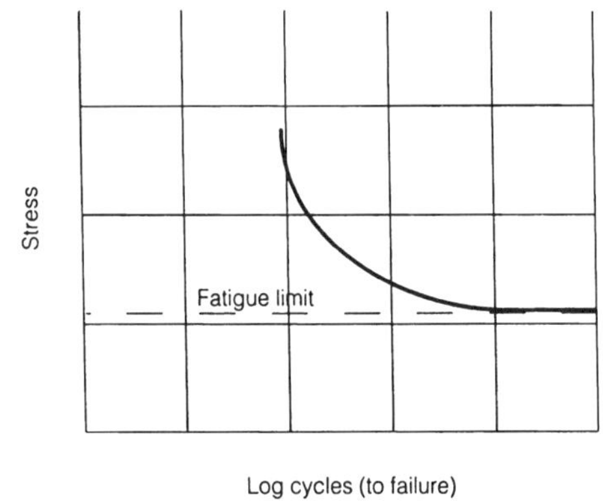

Fig. 1.19: Fatigue limit [6]

Some materials do not exhibit an asymptotic fatigue limit. In these cases, the endurance limit which gives stress or strain at failure at a certain number of cycles is used (Fig. 1.20) [6].

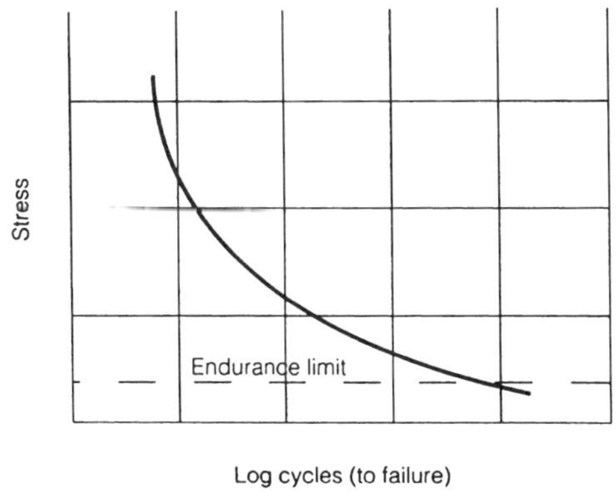

Fig. 1.20: Endurance limit [6]

For most plastics the fatigue limit is about 20 to 30 % of the ultimate strength measured in short-term tensile investigations [6]. The Woehler plots for oscillating flexural stress for some thermoplastics are given in Fig. 1.21. Fatigue limits decrease with increasing temperature, increasing frequency and stress concentrations in the part [4].

1.6 Hardness

Various methods of measuring the hardness of plastics are in use. Their common feature is measuring the deformation in terms of the depth of penetration which follows indentation by a hemisphere, cone or pyramid depending on the test procedure under defined load conditions. According to the type of indenter (Fig. 1.22) used the Shore hardness, for example, is given as Shore A or Shore D, Shore A data referring to soft plastics and Shore D to hard plastics. The results of both methods are expressed on a scale between 0 (very soft plastics) and 100 (very hard surface).

The ball-indentation hardness which represents the indentation depth of a spherical steel indenture is given in Table 1.3 for some thermoplastics [2].

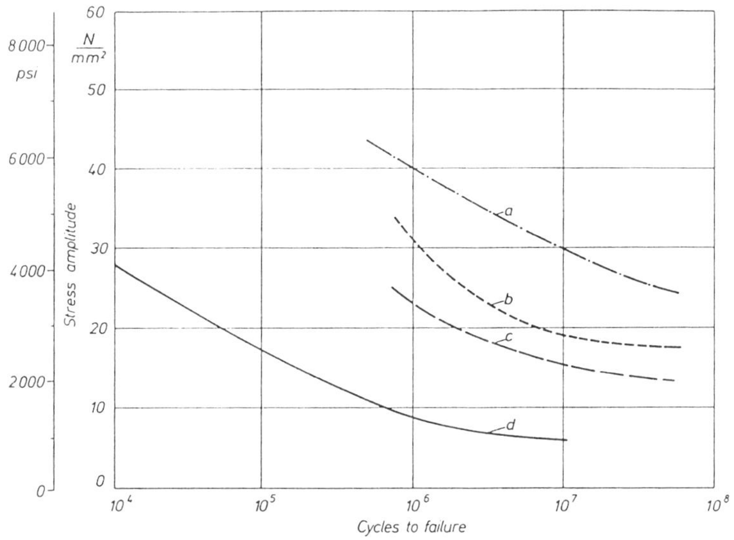

Fig. 1.21: Flexural fatigue strength of some thermoplastics [4]
a: acetal polymer, b: PP, c: PE-HD, d: PVC-U

Table 1.3: Ball-Indentation Hardness for Some Plastics [7]

Material	Hardness N/mm^2
PE-LD	13/20
PE-HD	40/65
PP	36/70
PVC-U	75/155
PS	120/130
ABS	80/120
PC	90/110
POM	150/170
PA6	70/75
PA66	90/100
PMMA	180/200
PET	180/200
PBT	150/180
Phenol-Formaldehyde-Resins	250 -320
Urea-Formaldehyde-Resins	260/350
Melamin-Formaldehyde-Resins	260/410
Unsaturated Polyester Resins	200/240

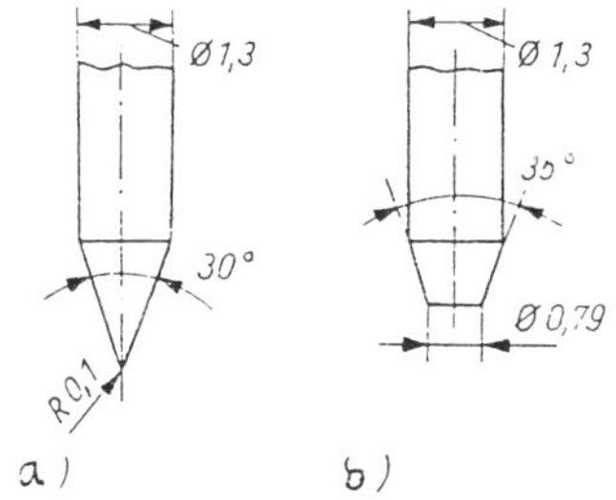

Fig. 1.22: Types of indenter [7]
a: shore, b: shore C and A

1.7 Impact Strength

Impact strength is the ability of the material to withstand a sudden impact blow as in a pendulum test, and indicates the toughness of the material at high rates of deformation. The test procedures are varied. In the Charpy impact test (Fig. 1.23) [2] the pendulum strikes the specimen centrally leading to fracture.

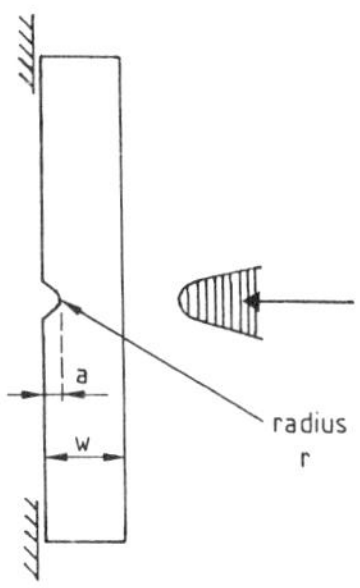

Fig. 1.23: Charpy impact test [2]

1.8 Coefficient of Friction

Although there is no consistent relationship between friction and wear, the factors affecting the two processes such as roughness of the surfaces, relative velocities of the parts in contact and area under pressure are often the same. Table 1.4 shows the coefficients of sliding friction and sliding wear against steel for different materials [4]. In applications where low friction and high wear resistance are required, smooth surface and low coefficient of friction of the resin components involved are to be recommended.

A typical impact curve as a function of temperature is shown in Fig. 1.24 [6]. This type of data provides the designer with information about the temperature at which the ductile fracture changes to brittle fracture thus enabling to evaluate the performance of the material in a given application. The results of impact tests depend on the manufacturing conditions of the specimen, notch geometry and on the test method. Impact strengths for various plastics are given by Domininghaus [4]. Charpy notched impact strengths of some plastics as functions of notch radius are given in Fig. 1.25 [5].

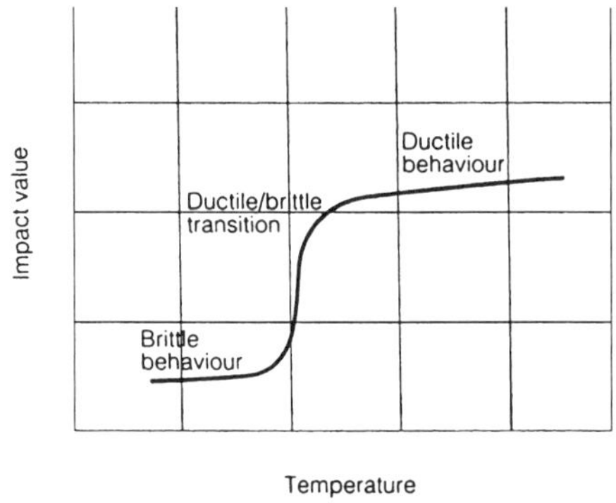

Fig. 1.24: Impact curve [6]

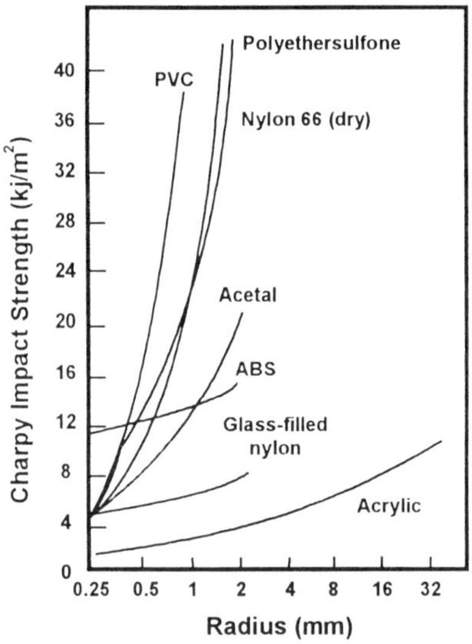

Fig. 1.25: Charpy impact strength vs. notch radius for some engineering thermoplastics [5]

Table 1.4: Coefficient of Sliding Friction of Polymer/Case Hardened Steel after 5 or 24 Hours [4]

Polymer	Coefficient of sliding friction	Sliding frictional wear
Polyamide 66	0.25/0.42	0.09
Polyamide 6	0.38/0.45	0.23
Polyamide 6 (*in situ* polymer)	0.36/0.43	0.10
Polyamide 610	0.36/0.44	0.32
Polyamide 11	0.32/0.38	0.8
Polyethyleneterephthalate	0.54	0.5
Acetal homopolymer	0.34	4.5
Acetal copolymer	0.32	8.9
Polypropylene	0.30	11.0
PE-HD (high molecular weight)	0.29	1.0
PE-HD (low molecular weight)	0.25	4.6
PE-LD	0.58	7.4
Polytetrafluoroethylene	0.22	21.0
PA 66 + 8% PE-LD	0.19	0.10
Polyacetal + PTFE	0.21	0.16
PA 66 + 3% MoS_2	0.32/0.35	0.7
PA 66 – GF 35	0.32/0.36	0.16
PA 6 – GF 35	0.30/0.35	0.28
Standard polystyrene	0.46	11.5
Styrene/Acrylonitrile copolymer	0.52	23
Polymethylmethacrylate	0.54	4.8
Polyphenylene ether	0.35	90

Example [8]:
The following example illustrates the use of physical and performance properties of polymers in dealing with design problems.

The minimum depth of the simple beam of SAN shown in Fig. 1.26 is to be determined for the following conditions:

The beam should support a mid-span load of 11.13 N for 5 years without fracture and without causing a deflection of greater than 2.54 mm.

Solution:
The maximum stress is given by

$$\sigma_{max} = \frac{1.5 \; F \cdot l}{b \cdot d^2} \tag{1.8.1}$$

where F = load (N)

l, b, d = dimensions in (mm) as shown in Fig. 1.26.

The creep modulus E_c is calculated from

$$E_c = \frac{F \cdot l^3}{4 \cdot f \cdot b \cdot d^3} \tag{1.8.2}$$

where f is deflection in mm.

The maximum stress from Fig. 1.27 at a period of 5 years (= 43800 h) is

$$\sigma_{max} = 23.44 \ N/mm^2 .$$

Working stress σ_w with an assumed safety factor S = 0.5:

$$\sigma_w = 23.44 \cdot 0.5 = 11.72 \ N/mm^2 .$$

Creep modulus E_c at $\sigma < \sigma_w$ and a period of 5 years from Fig. 1.27:

$$E_c = 2413 \ N/mm^2 .$$

Creep modulus with a safety factor S = 0.75:

$$E_c = 2413 \cdot 0.75 = 1809.75 \ N/mm^2 .$$

The depth of the beam results from Eq. (1.8.1)

$$d = \left(\frac{1.5 \cdot F \cdot l}{b \cdot \sigma_{max}} \right)^{0.5} = \left(\frac{1.5 \cdot 11.13 \cdot 76.2}{12.7 \cdot 11.72} \right)^{0.5} = 2.92 \text{ mm}.$$

The deflection is calculated from Eq. (1.8.2)

$$f = \frac{F \cdot l^3}{4 \cdot E_C \cdot b \cdot d^3} = \frac{11.13 \cdot 76.2^3}{4 \cdot 1809.75 \cdot 12.7 \cdot 2.92^3} = 2.15 \text{ mm}.$$

Under these assumptions the calculated f is less than the allowable value of 2.54 mm.

This example shows why creep data are required in design calculations and how they can be applied to solve design problems.

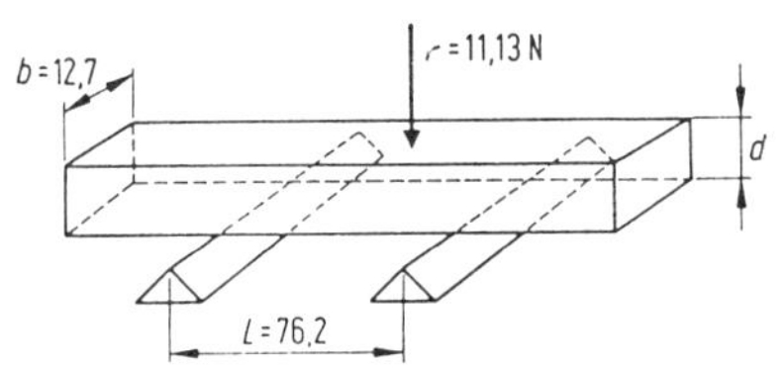

Fig. 1.26: Beam under mid-span load [8]

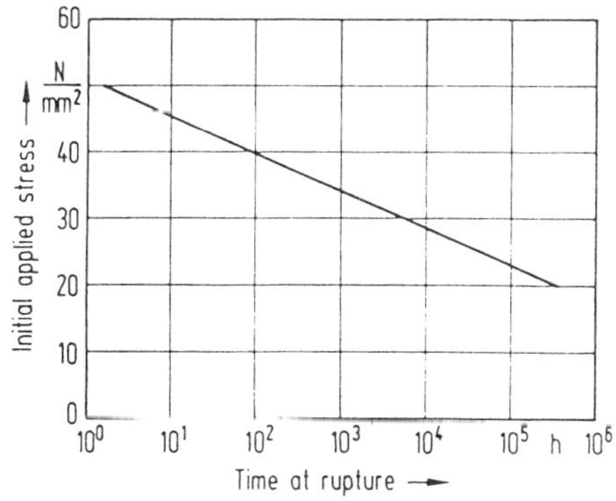

Fig. 1.27: Creep curve of SAN [8]

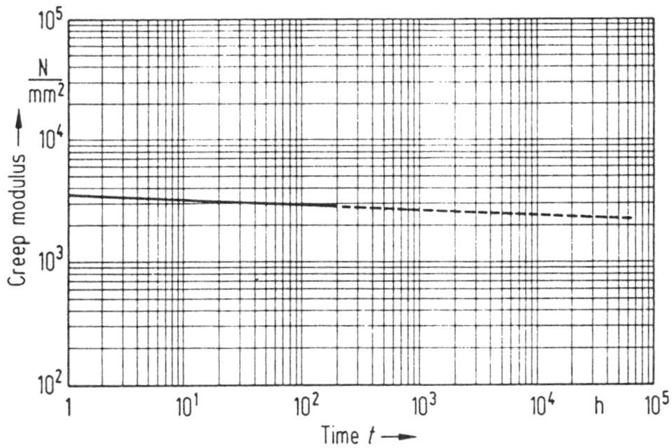

Fig. 1.28: Creep modulus of SAN [8]

Literature

1. *Nat, D.S.:* A Text Book of Materials and Metallurgy, Katson Publishing House, Ludhiana, India 1980
2. *Birley, A.W.: Haworth, B.: Bachelor, T.:* Physics of Plastics, Hanser, Munich 1991
3. *Rao, N.S.:* Design Formulas for Plastics Engineers, Hanser, Munich, 1991
4. *Domininghaus, H.:* Plastics for Engineers, Hanser, Munich 1993
5. *Rigby, R.B.:* Polyethersulfone in Engineering Thermoplastics: Properties and Applications. Ed.: James M. Margolis. Marcel Dekker, Basel 1985
6. *General Electric Plastics Brochure:* Engineering Materials Design Guide
7. *Schmiedel, H.:* Handbuch der Kunststoffprüfung (Ed.), Hanser, Munich 1992

8. *Design Guide, Modern Plastics Encyclopedia*, 1978-79
9. *Brochure: Advanced CAE Technology Inc. 1992*
10. *Pahl, M.; Gleißle, W.; Laun, H.M.:* Praktische Rheologie der Kunststoffe und Elastomere, VDI-Kunststofftechnik, Düsseldorf 1995

2 Thermal Properties of Solid and Molten Polymers

In addition to the mechanical and melt flow properties thermodynamic data of polymers are necessary for optimizing various heating and cooling processes which occur in plastics processing operations.

In design work the thermal properties are often required as functions of temperature and pressure [2]. As the measured data cannot always be predicted by physical relationships accurately enough, regression equations are used to fit the data for use in design calculations.

2.1 Specific Volume

The volume-temperature relationship as a function of pressure is shown for a semi-crystalline PE-HD in Fig. 2.1 [1] and for an amorphous PS in Fig. 2.2 [1]. The p-v-T diagrams are needed in many applications, for example to estimate the shrinkage of plastics parts in injection molding [19]. Data on p-v-T relationships for a number of polymers are presented in the handbook [8].

According to Spencer-Gilmore equation which is similar to the Van-der-Waal equation of state for real gases the relationship between pressure p, specific volume v and temperature T of a polymer can be written as

$$(v - b^*)\,(p + p^*) = \frac{RT}{W} \tag{2.1.1}$$

In this equation b^* is the specific individual volume of the macromolecule, p^*, the cohesion pressure, W, the molecular weight of the monomer and R, the universal gas constant [9].

The values p^* and b^* can be determined from p-V-T diagrams by means of regression analysis. Spencer and Gilmore and other workers evaluated these constants from measurements for the polymers listed in Table 2.1 [9], [18].

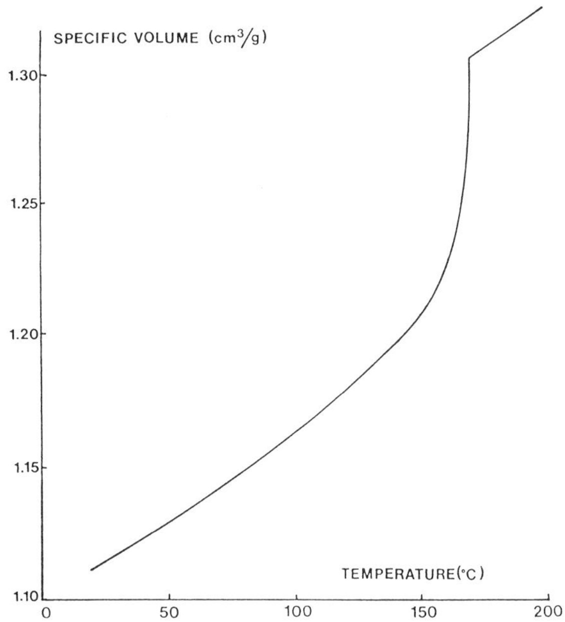

Fig. 2.1: Specific volume vs. temperature for a semi-crystalline polymer (PP) [1]

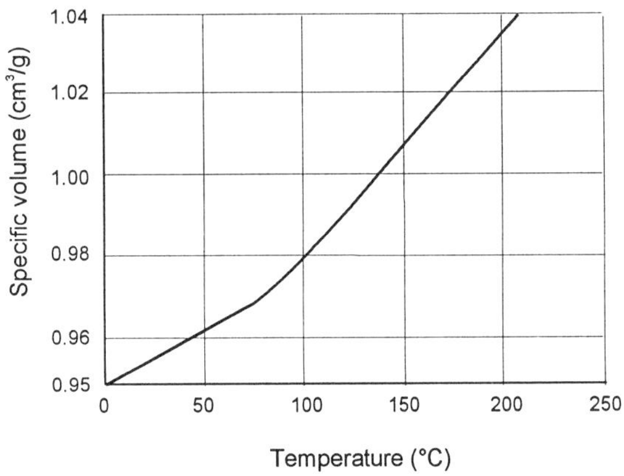

Fig. 2.2: Specific volume vs. temperature for an amorphous polymer (PS) [1]

Table 2.1: Constants for the Equation of State [9]

Material	W g/mol	p^* atm	b^* cm^3/g
PE-LD	28.1	3240	0.875
PP	41.0	1600	0.620
PS	104	1840	0.822
PC	56.1	3135	0.669
PA 610	111	10768	0.9064
PMMA	100	1840	0.822
PET	37.0	4275	0.574
PBT	113.2	2239	0.712

Example:
Following values are given for a PE-LD:

$$W = 28.1 \text{ g/Mol}$$
$$b^* = 0.875 \text{ cm}^3/\text{g}$$
$$p^* = 3240 \text{ atm}$$

Calculate the specific volume at

$$T = 190\ °C \text{ and } p = 1 \text{ bar}$$

Solution:
Using Eq. (2.1.1) and the conversion factors to obtain the volume v in cm^3/g, we obtain

$$v = \frac{10 \cdot 8.314 \cdot (273 + 190)}{28.1 \cdot 3240.99 \cdot 1.013} + 0.875 = 1.292 \ \text{cm}^3/\text{g}$$

The density ρ is the reciprocal value of specific volume so that

$$\rho = \frac{1}{v}$$

The p-v-T data can also be fitted by a polynomial of the form

$$v = A(0)_v + A(1)_v \cdot p + A(2)_v \cdot T + A(3)_v \cdot T \cdot p \tag{2.1.2}$$

if measured data is available (Fig. 2.3) [10], [11], [17]. The empirical coefficients $A(0)_v$ ... $A(3)_v$ can be determined by means of the computer program given in [11]. With the modified two-domain Tait equation [16] a very accurate fit can be obtained both for the solid and melt regions.

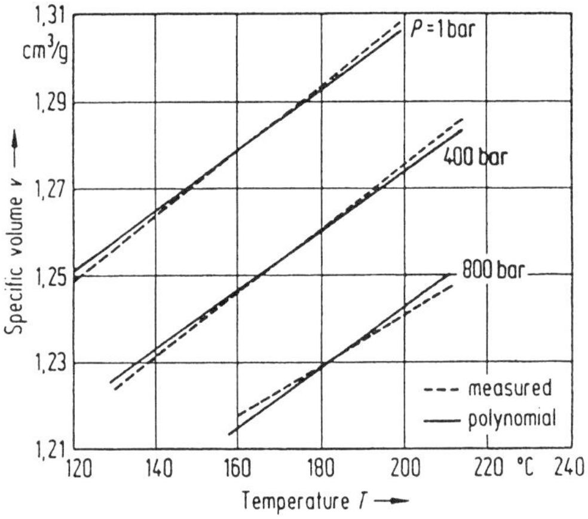

Fig. 2.3: Specific volume as a function of temperature and pressure for PE-LD [2], [13]

2.2 Specific Heat

The specific heat c_p is defined as

$$c_p = \left(\frac{\partial h}{\partial T}\right)_p \tag{2.2.1}$$

where h = enthalpy
T = Temperature

The specific heat c_p gives the amount of heat which is supplied to a system in a reversible process at a constant pressure in order to increase the temperature of the substance by dT. The specific heat at constant volume c_v is given by

$$c_v = \left(\frac{\partial u}{\partial T}\right)_v \tag{2.2.2}$$

where u = internal energy
T = Temperature

In the case of c_v the supply of heat to the system occurs at constant volume.

c_p and c_v are related to each other through the Spencer-Gilmore equation Eq. (2.1.1)

$$c_v = c_p - \frac{R}{W} \tag{2.2.3}$$

The numerical values of c_p and c_v differ by roughly 10 %, so that for approximate calculations c_v can be made equal to c_p.

Plots of c_p as function of temperature are shown in Fig. 2.4 for amorphous, semi-crystalline and crystalline polymers.

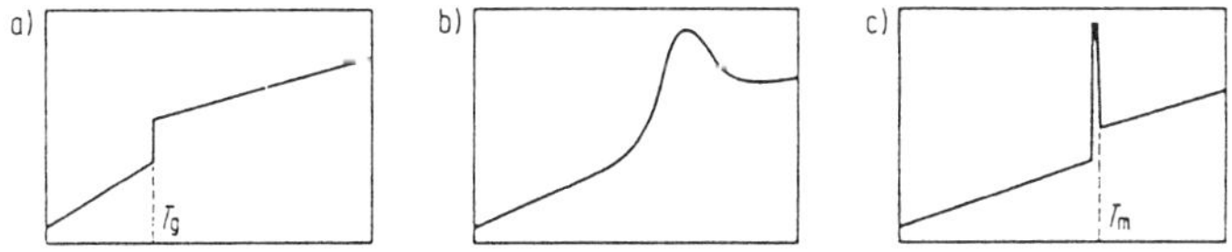

Fig. 2.4: Specific heat as a function of temperature for amorphous (a), semi-crystalline (b) and crystalline polymers (c) [14]

As shown in Fig. 2.5 measured values can be fitted by a polynomial of the type [11]

$$c_p(T) = A(0)\,c_p + A(1)\,c_p \cdot T + A(2)\,c_p \cdot T^2 \tag{2.2.4}$$

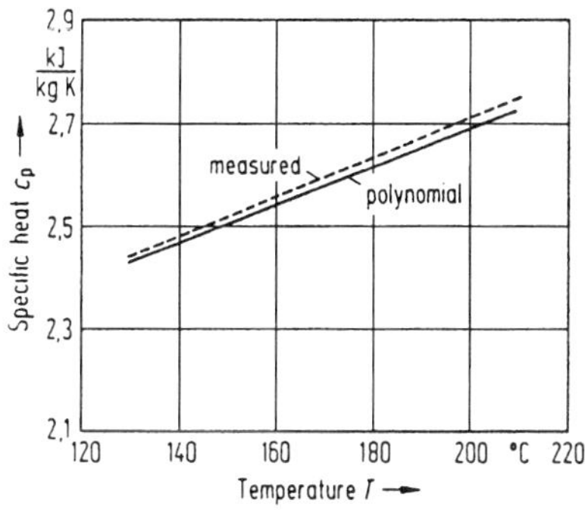

Fig. 2.5: Comparison between measured values of c_p [13] and polynomial for PE-LD [2]

2.3 Thermal Expansion Coefficient

The expansion coefficient α_v at constant pressure is given by [14]

$$\alpha_v = \frac{1}{v}\left(\frac{\partial v}{\partial T}\right)_p \tag{2.3.1}$$

The isothermal compression coefficient γ_k is defined as [14]

$$\gamma_k = -\frac{1}{v}\left(\frac{\partial v}{\partial p}\right)_T \qquad (2.3.2)$$

α_v and γ_k are related to each other by the expression [14]

$$c_p = c_v + \frac{T \cdot v \cdot \alpha_v^2}{\gamma_k} \qquad (2.3.3)$$

The linear expansion coefficient α_{lin} is approximately

$$\alpha_{lin} = \frac{1}{3}\alpha_v \qquad (2.3.4)$$

Table 2.2 shows the linear expansion coefficients of some polymers at 20 °C. The linear expansion coefficient of mild steel lies around 11×10^{-6} [K^{-1}] and that of aluminum about 25×10^{-6} [K^{-1}]. As can be seen from Table 2.2 plastics expand about 3 to 20 times more than metals. Factors affecting thermal expansion are crystallinity, cross-linking and fillers [1].

Table 2.2: Coefficients of Linear Thermal Expansion [1], [5]

Polymer	Coefficient of linear expansionat 20 °C α_{lin} $10^6 K^{-1}$
PE-LD	250
PE-HD	200
PP	150
PVC-U	75
PVC-P	180
PS	70
ABS	90
PMMA	70
POM	100
PSU	50
PC	65
PET	65
PBT	70
PA 6	80
PA 66	80
PTFE	100
TPU	150

2.4 Enthalpy

Eq. (2.2.1) leads to

$$dh = c_p \cdot dT \tag{2.4.1}$$

As shown in Fig. 2.6 the measured data on h = h(T) [13] for a polymer melt can be fitted by the polynomial

$$h(T) = A(0)_h + A(1)_h \cdot T + A(2)_h \cdot T^2 \tag{2.4.2}$$

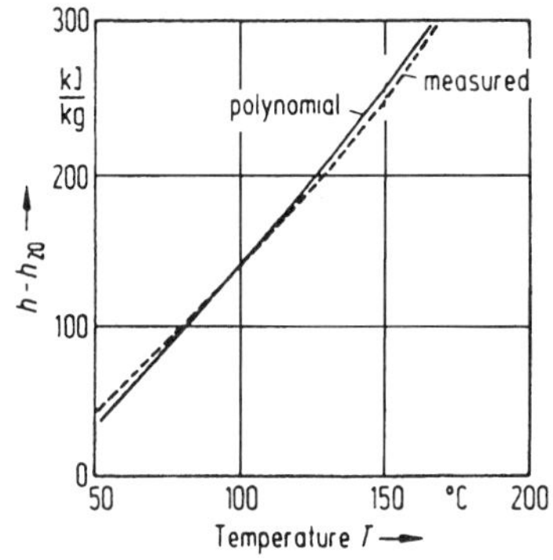

Fig. 2.6: Comparison between measured values of h [13] and polynomial for PA6 [11]

The specific enthalpy defined as the total energy supplied to the polymer divided by the throughput of the polymer is a useful parameter for designing extrusion and injection molding equipment such as screws. It gives the theoretical amount of energy required to bring the solid polymer to the process temperature. Values of this parameter for different polymers are given in Fig. 2.7 [14].

If, for example, the throughput of an extruder is 100 kg/h of polyamide (PA) and the processing temperature is 260° C, the theoretical power requirement would be 20 kW. This can be assumed to be a safe design value for the motor horse power, although theoretically it includes the power supply to the polymer by the heater bands of the extruder as well.

2.5 Thermal Conductivity

The thermal conductivity λ is defined as

$$\lambda = \frac{Q \cdot l}{t \cdot A \cdot (T_1 - T_2)} \tag{2.5.1}$$

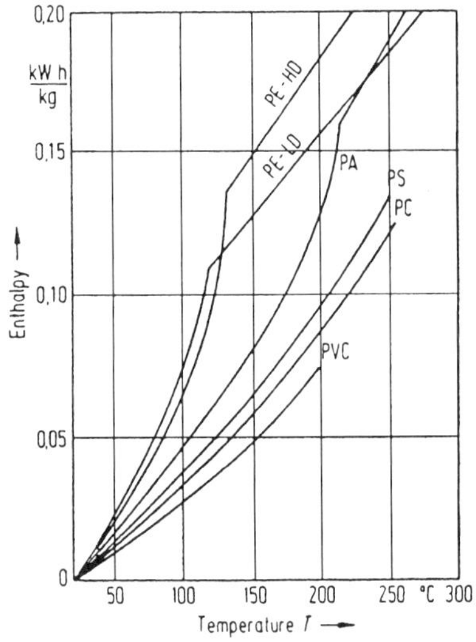

Fig. 2.7: Specific enthalpy as a function of temperature [14]

where Q = heat flow through the surface of area A in a period of time t

(T_1-T_2) = temperature difference over the length l.

Analogous to the specific heat c_p and enthalpy h the thermal conductivity λ as shown in Fig. 2.8 can be expressed as [2]

$$\lambda(T) = A(0)_\lambda + A(1)_\lambda \cdot T + A(2)_\lambda \cdot T^2 \qquad (2.5.2)$$

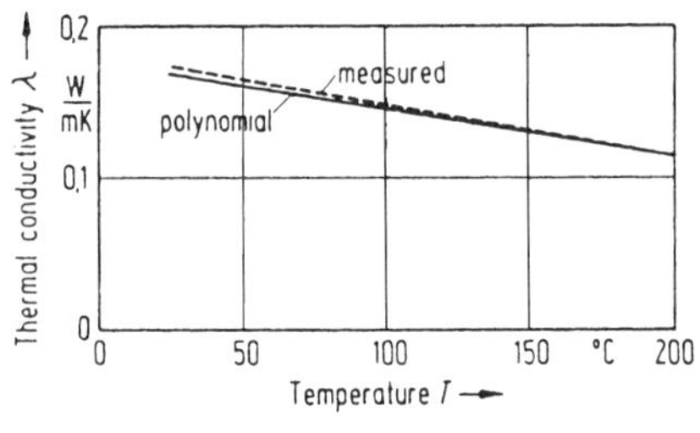

Fig. 2.8: Comparison between measured values of λ [13] and polynomial for PP [2]

The thermal conductivity increases only slightly with the pressure. A pressure increase from 1 bar to 250 bar leads only to an increase of less than 5 % of its value at 1 bar.

As in the case of other thermal properties the thermal conductivity is, in addition to its dependence on temperature, strongly influenced by the crystallinity and orientation and by the amount and type of filler in the polymer [1]. Foamed plastics have, for example, thermal conductivities at least an order of magnitude less than those of solid polymers [1].

2.6 Thermal Diffusivity

Thermal diffusivity a is defined as the ratio of thermal conductivity to the heat capacity per unit volume [1]

$$a = \frac{\lambda}{\rho \cdot c_p} \tag{2.6.1}$$

and is of importance in dealing with transient heat transfer phenomena such as cooling of melt in an injection mold [2]. Although for approximate calculations average values of thermal diffusivity can be used, more accurate computations require functions of λ, ρ and c_p against temperature for the solid as well as melt regions of the polymer. Thermal diffusivities of some polymers at 20 °C are listed in Table 2.3 [14], [15].

Table 2.3: Thermal Diffusivities of Polymers at 20 °C

Polymer	Thermal diffusivity at 20 °C a 10^6 m^2/s
PE-LD	0.12
PE-HD	0.22
PP	0.14
PVC-U	0.12
PVC-P	0.14
PS	0.12
PMMA	0.12
POM	0.16
ABS	0.15
PC	0.13
PBT	0.12
PA 6	0.14
PA 66	0.12
PET	0.11

Exhaustive measured data of the quantities c_p, h, λ and pvT diagrams of polymers are given in the VDMA-Handbook [8]. Approximate values of thermal properties of use to plastics engineers are summarized in Table 2.4 [14], [15].

Experimental techniques of measuring enthalpy, specific heat, melting point and glass transition temperature by differential thermal analysis (DTA) such as differential scanning calorimetry (DSC) are described in detail in the brochure [12]. Methods of determining thermal conductivity, pvT values and other thermal properties of plastics have been treated in this brochure [12].

Table 2.4: Approximate Values for Thermal Properties of Some Polymers [14]

Polymer	Thermal conductivity (20 °C) λ W/m·K	Specific heat (20 °C) c_p kJ/kgK	Density (20 °C) ρ g/cm³	Glass transition temperature T_g °C	Melting point range T_m °C
PS	0.12	1.20	1.06	101	-
PVC	0.16	1.10	1.40	80	-
PMMA	0.20	1.45	1.18	105	-
SAN	0.12	1.40	1.08	115	-
POM	0.25	1.46	1.42	-73	ca. 175
ABS	0.15	1.40	1.02	115	-
PC	0.23	1.17	1.20	150	-
PE-LD	0.32	2.30	0.92	-120/-90	ca. 110
PE-LLD	0.40	2.30	0.92	-120/-90	ca. 125
PE-HD	0.49	2.25	0.95	-120/-90	ca. 130
PP	0.15	2.40	0.91	- 10	160/170
PA6	0.36	1.70	1.13	50	215/225
PA 66	0.37	1.80	1.14	55	250/260
PET	0.29	1.55	1.35	70	250/260
PBT	0.21	1.25	1.35	45	ca. 220

2.7 Coefficient of Heat Penetration

The coefficient of heat penetration is used in calculating the contact temperature which results when two bodies of different temperature are brought into contact with each other [2], [16].

As shown in Table 2.5 the coefficients of heat penetration of metals are much higher than those of polymer melts. Owing to this the contact temperature of the wall of an injection mold at the time of injection lies in the vicinity of the mold wall temperature before injection.

The contact temperature $\theta_{w\,max}$ of the wall of an injection mold at the time of injection is [17]

$$\theta_{w\,max} = \frac{b_w\ \theta_{w\,min} + b_p\ \theta_M}{b_w + b_p} \tag{2.7.1}$$

where b = coefficient of heat penetration = $\sqrt{\lambda\ \rho\ c}$

$\theta_{w\,min}$ = temperature before injection

θ_M = melt temperature

Indices w and p refer to mold and polymer respectively.

Table 2.5: Coefficients of Heat Penetration of Metals and Plastics [17]

Material	Coefficient of heat penetration b $W \cdot s^{0.5} \cdot m^{-2} \cdot K^{-1}$
Beryllium Copper (BeCu25)	17.2×10^3
Unalloyed Steel (C45W3)	13.8×10^3
Chromium Steel (X40Cr13)	11.7×10^3
Polyethylene (PE-HD)	0.99×10^3
Polystyrene (PS)	0.57×10^3
Stainless Steel	7.56×10^3
Aluminum	21.8×10^3

The values given in Table 2.5 refer to the following units of the properties:

Thermal conductivity λ: W/(m·K)
Density ρ: kg/m^3
Specific heat c: kJ/(kg·K)

The approximate values for steel are

l = 50 W/(m·K)
r = 7850 kg/m^3
c = 0.485 kJ/(kg·K)

The coefficient of heat penetration is

$$b = \sqrt{\lambda \cdot \rho \cdot c} = \sqrt{50 \cdot 7.85 \cdot 10^3 \cdot 0.485 \cdot 10^3} = 13.8 \cdot 10^3\ W\,s^{0.5} \cdot m^{-2} \cdot K^{-1}$$

2.8 Heat Deflection Temperature

Heat Deflection Temperature (HDT) or the Deflection Temperature under load (DTUL) is a relative measure of the polymer's ability to retain shape at elevated temperatures while supporting a load. In amorphous materials the HDT almost coincides with the glass transition temperature Tg. Crystalline polymers may have lower values of HDT but are dimensionally more stable at elevated temperatures [5]. Additives like fillers have more significant effect on crystalline polymers than on amorphous polymers. The heat deflection temperatures are listed for some materials in Table 2.6. Owing to the similarity of the measuring principle Vicat softening point, HDT and Martens temperature lie often close to each other (Fig. 2.9) [6].

Table 2.6: Heat Deflection Temperatures (HDT) According to the Method A of Measurement [3], [6]

Material	HDT (Method A) °C
PE-LD	35
PE-HD	50
PP	45
PVC	72
PS	84
ABS	100
PC	135
POM	140
PA6	77
PA66	130
PMMA	103
PET	80
PBT	65

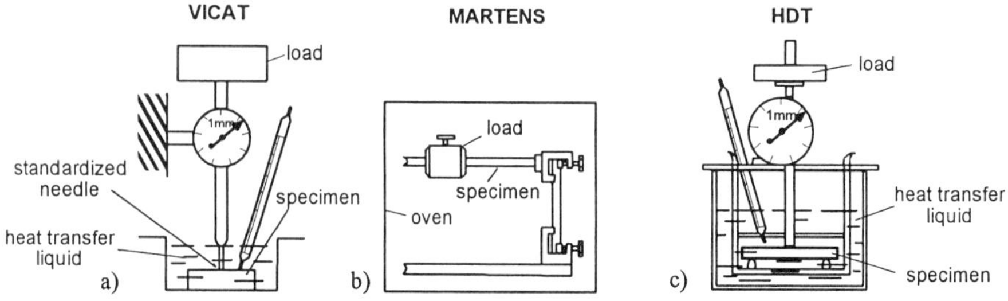

Fig. 2.9: Principles of measurement of heat distortion of plastics [6]

2.9 Vicat Softening Point

Vicat softening point is the temperature at which a small lightly loaded, heated test probe penetrates a given distance into a test specimen [5].

The Vicat softening point gives an indication of the material's ability to withstand contact with a heated object for a short duration. It is used as a guide value for demolding temperature in injection molding. Vicat softening points of crystalline polymers have more significance than amorphous polymers, as the latter tend to creep during the test [5]. Both Vicat and HDT values serve as a basis to judge the resistance of a thermoplastic to distortion at elevated temperatures.

Guide values of Vicat softening temperatures of some polymers according to DIN 53460 (Vicat 5 kg) are given in Table 2.7 [6].

Table 2.7: Guide Values for Vicat Softening Points [6]

Polymer	Vicat softening point °C
PE-HD	65
PP	90
PVC	92
PS	90
ABS	102
PC	138
POM	165
PA 6	180
PA 66	200
PMMA	85
PET	190
PBT	180

2.10 Flammability

Most plastics are flammable and can act as fuel in a fire situation. The aim of various flammability tests is to measure the burning characteristics of plastics materials once ignition has occurred. The Critical Oxygen Index is the most accepted laboratory test. It measures the minimum volume concentration of oxygen in a mixture of oxygen and nitrogen, which is required to initiate combustion of the plastic and propagate flame-burning to a pre-specified extent (a flame-spread time or a distance along the test sample) [1]. Data are presented in the form of Limiting Oxygen Index (LOI) value. These values are summarized in Table 2.8 [1]. The non-flammability of PTFE is manifested by its exceptionally high LOI value. Other flame-resistant plastics include thermosets, PVC and PVDC. Resistance to flammability can be enhanced by adding flame retarding additives to the polymer [1].

Table 2.8: Limiting Oxygen Indices (LOI) of polymers (23 °C) [1]

Material	LOI
POM	14.9/15.7
PMMA	17.4
PE	17.4
PP	17.4
PS	17.6/18.3
ABS	18.3/18.8
PET	20.0
PA 66	24/29
PSU	30/32
PI	36.5
PVDC	60.0
PVC (chlorinated)	60.0/70.0
PTFE	95.0

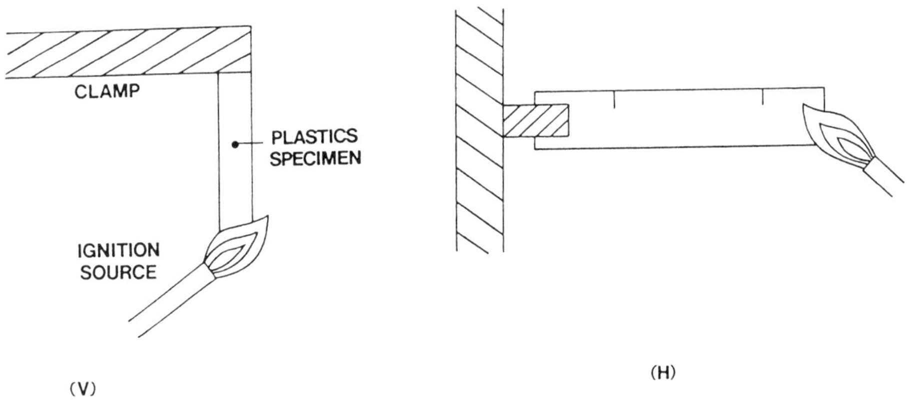

Fig. 2.10: Typical arrangements for the UL flammability test: (V) UL 94 v vertical small flame ignitibility (variable sample thickness); (H) UL 94 HB horizontal small flame ignitibility / flame spread [1]

Underwriters' Laboratory test (UL 94) is a rating system which classifies flammability behavior of plastics according to their ability to maintain combustion after the flame is removed (see Fig. 2.10 [1]). In general, plastics which extinguish rapidly and do not drip flaming particles obtain high ratings. Ratings are given on the basis of minimum wall thickness of the material which corresponds to a particular flame class [1]. These thicknesses are important for designing plastics component design.

The temperature index determined on the basis of the tensile half-life concept of the Underwriters' Laboratories is shown for various polymers in Fig. 2.11 [4]. The smoke emission for a number of plastics is given in Fig. 2.12 [4].

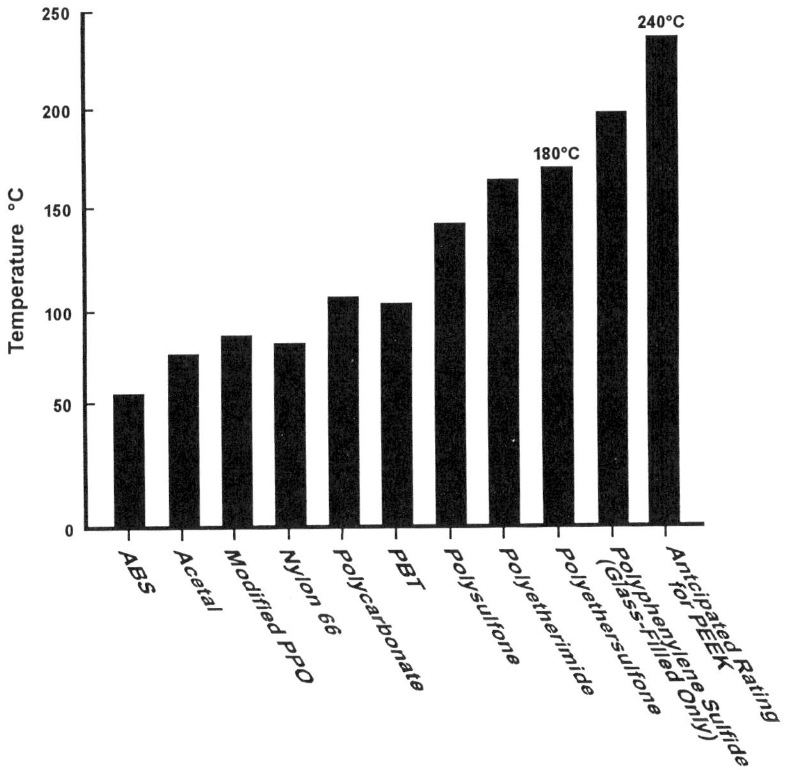

Fig. 2.11: UL temperature index for some thermoplastics [4]

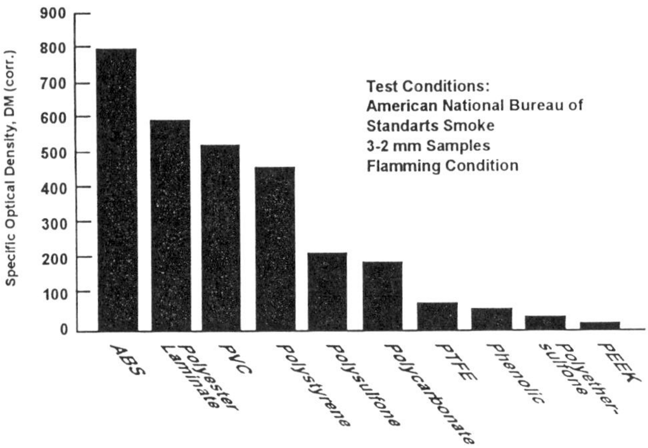

Fig. 2.12: Smoke emission behavior of some thermoplastics [4]

Literature

1. *Birley, A.W.: Haworth, B.: Bachelor, T.:* Physics of Plastics, Hanser, Munich 1991
2. *Rao, N.S.:* Design Formulas for Plastics Engineers, Hanser, Munich, 1991
3. *Domininghaus, H.:* Plastics for Engineers, Hanser, Munich 1993
4. *Rigby, R.B.:* Polyethersulfone in Engineering Thermoplastics: Properties and Applications. Ed.: James M. Margolis. Marcel Dekker, Basel 1985
5. *General Electric Plastics Brochure:* Engineering Materials Design Guide
6. *Schmiedel, H.:* Handbuch der Kunststoffprüfung (Ed.), Hanser, Munich 1992
7. *Design Guide, Modern Plastics Encyclopedia*, 1978-79
8. *Kenndaten für die Verarbeitung thermoplastischer Kunststoffe*, Teil I, Thermodynamik, Hanser, Munich 1979
9. *Mckelvey, J.M.:* Polymer Processing, John Wuley, New York 1962
10. *Münstedt, H.:* Berechnen von Extrudierwerkzeugen, VDI-Verlag, Düsseldorf, 1978
11. *Rao, N.S.:* Designing Machines and Dies Polymer Processing, Hanser, Munich 1981
12. *Brochure: Advanced CAE Technology Inc. 1992*
13. *Proceedings, 9. Kunststofftechnisches Kolloquium*, IKV, Aachen 1978
14. *Rauwendal, C.:* Polymer Extrusion, Hanser, Munich 1986.
15. *Ogorkiewicz, R.M.:* Thermoplastics Properties and Design, John Wuley, New York 1973
16. *Martin, H.:* VDI-Wärmeatlas, VDI-Verlag, Düsseldorf 1984
17. *Wübken, G.:* Berechnen von Spritzgießwerkzeugen, VDI-Verlag, Düsseldorf 1974
18. *Progelhof, R.C., Throne, J.L.:* Polymer Engineering Principles - Properties, Processes, Tests for Design, Hanser, Munich 1993
19. *Kurfess, W.:* Kunststoffe 61 (1971), p 421

3 Transport Properties of Molten Polymers

The basic principle of making parts out of polymeric materials lies in creating a melt from the solid material and forcing the melt into a die, the shape of which corresponds to that of the part. Thus, as Fig. 3.1 indicates, melt flow and thermal properties of polymers play an important role in the operations of polymer processing.

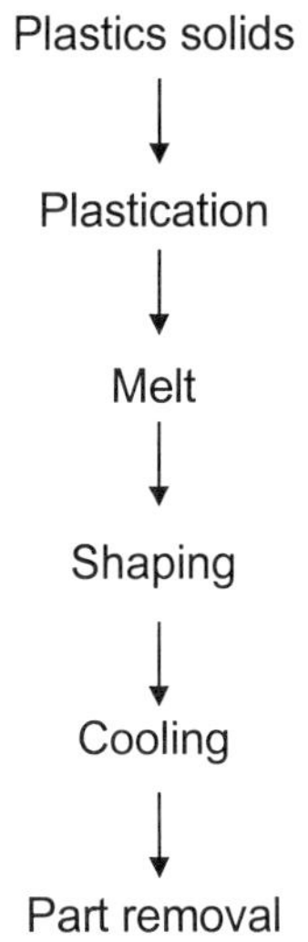

Fig. 3.1: Principle of manufacturing of plastics parts

3.1 Newtonian and Non-Newtonian Fluids

Analogous to the ideal elastic solids there exists a linear relationship between stress and strain in the case of Newtonian fluids (Fig. 3.2).

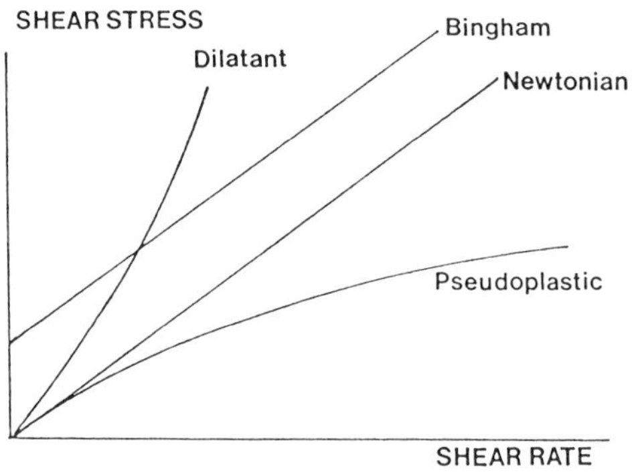

Fig. 3.2: Flow curves of idealized fluids [1]

The fluid between the upper plate in Fig. 3.3 moving at a constant velocity U_x and the lower stationary plate experiences a shear stress τ (see Fig. 1.1B also).

$$\tau = \frac{F_t}{A} \tag{3.1.1}$$

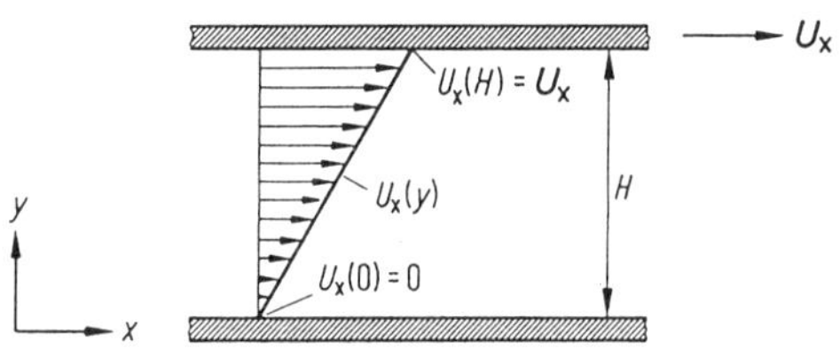

Fig. 3.3: Shear flow

The shear or deformation rate of the fluid is equal to

$$\dot{\gamma} = \frac{U_x}{H} = \frac{du}{dy} \tag{3.1.2}$$

The shear viscosity is then defined as

$$\eta = \frac{\tau}{\dot{\gamma}} \tag{3.1.3}$$

For an extensional flow, which corresponds to the tension test of Hookean solid, we get

$$\sigma_Z = \lambda \cdot \dot{\varepsilon} \tag{3.1.4}$$

where

σ_Z = normal stress
λ = Trouton viscosity
$\dot{\varepsilon}$ = strain rate

Analogous to Eq. (1.4.4) one obtains

$$\lambda = 3\eta \tag{3.1.5}$$

3.2 Viscous Shear Flow

Macromolecular fluids like thermoplastic melts exhibit significant non-Newtonian behavior. This is noticed in the marked decrease of melt viscosity when the melt is subjected to shear or tension as shown in Fig. 3.4 and in Fig. 3.5. The flow of melt in the channels of dies and polymer processing machinery is mainly shear flow. Therefore, knowledge of the laws of shear flow is necessary for designing machines and dies for polymer processing. For practical applications the following summary of the relationships was found to be useful.

3.2.1 Apparent Shear Rate

The apparent shear rate for a melt flowing through a capillary is defined as

$$\dot{\gamma}_a = \frac{4\dot{Q}}{\pi R^3} \tag{3.2.1}$$

where $\dot{Q}$ is the volume flow rate per second and R, the radius of capillary.

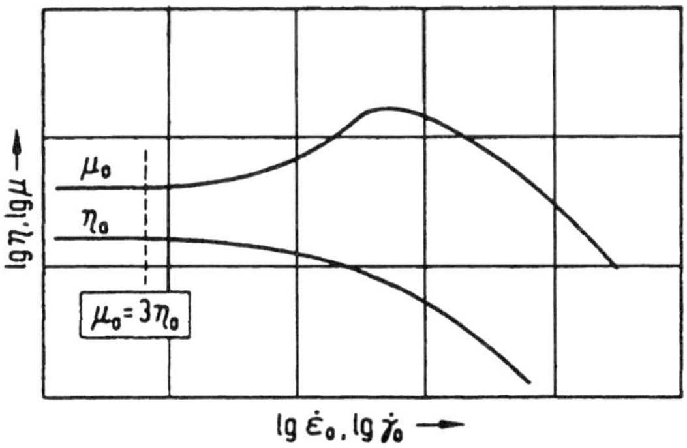

Fig. 3.4: Tensile viscosity and shear viscosity of a polymer melt as a function of strain rate [14]

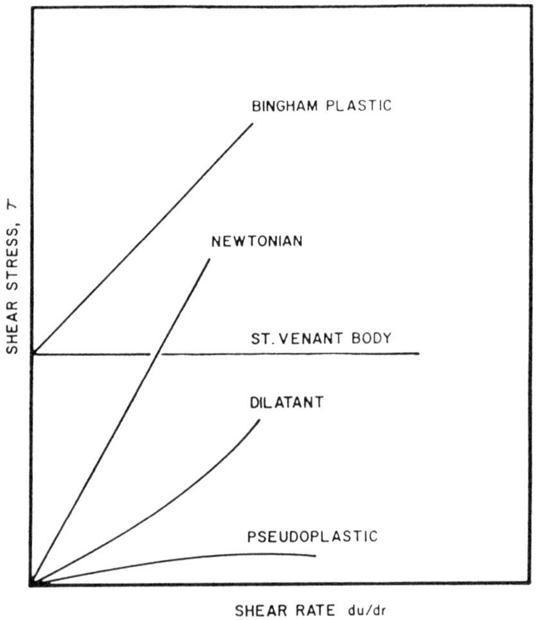

Fig. 3.5: Shear stress as a function of shear rate for different types of plastics [24]

3.2.2 Entrance Loss

Another rheological parameter which is of practical importance is the entrance loss p_c representing the loss of energy of flow at the entrance to a round nozzle. This is correlated empirically by the relation [4]

$$p_c = c \cdot \tau^m \tag{3.2.2}$$

where c and m are empirical constants and τ the shear stress. These constants can be determined from the well-known Bagley curves as shown in Fig. 3.6. The values of these constants are given in Table 3.1 for some of the thermoplastics. The dimensions of shear stress and entrance loss used in the calculation of c and m are Pa.

3.2.3 True Shear Stress

The flow curves of a particular PE-LD measured with a capillary rheometer are given in Fig. 3.7. The plot shows the apparent shear rate $\dot{\gamma}_a$ as a function of the true shear stress τ

at the capillary wall with the melt temperature as a parameter. The entrance loss p_c was obtained from the Bagley plot shown in Fig. 3.6.

Thus, the true shear stress τ is given by

$$\tau = \frac{p - p_c}{2(L/R)} \tag{3.2.3}$$

where L = length of the capillary

R = radius of the capillary respectively.

p = pressure of the melt (see Fig. 3.6)

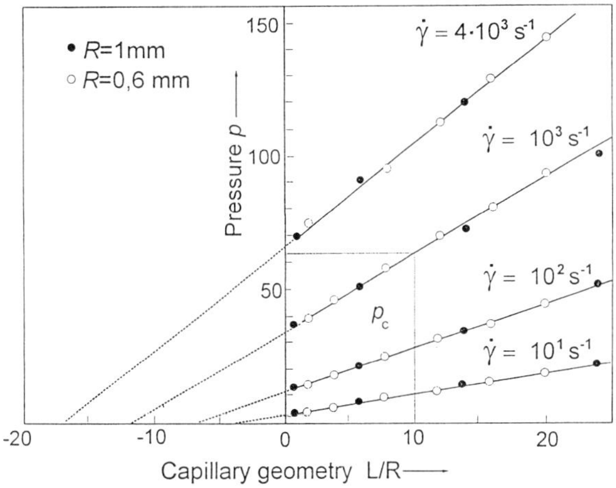

Fig. 3.6: Bagley plots of a polystyrene with the capillary length L and radius R [15]

Table 3.1: Resin-Dependent Constants c and m in Eq. (3.2.2) [4]

Product	c	m
Polypropylene (Novolen 1120 H)	$2.551 \cdot 10^{-5}$	2.116
Polypropylene (Novolen 1120 L)	$1.463 \cdot 10^{-4}$	1.976
Polypropylene (Novolen 1320 L)	$2.871 \cdot 10^{-7}$	2.469
PE-LD (Lupolen 1800 M)	$1.176 \cdot 10^{-1}$	1.434
PE-LD (Lupolen 1800 S)	$6.984 \cdot 10^{0}$	1.072
PE-LD (Lupolen 1810 D)	$5.688 \cdot 10^{-4}$	1.905
PE-HD (Lupolen 6011 L)	$3.940 \cdot 10^{-2}$	1.399
PE-HD (Lupolen 6041 D)	$1.788 \cdot 10^{0}$	1.187
Polyisobutylene (Oppanol B10)	$6.401 \cdot 10^{-3}$	1.575
Polyisobutylene (Oppanol B15)	$1.021 \cdot 10^{-7}$	2.614

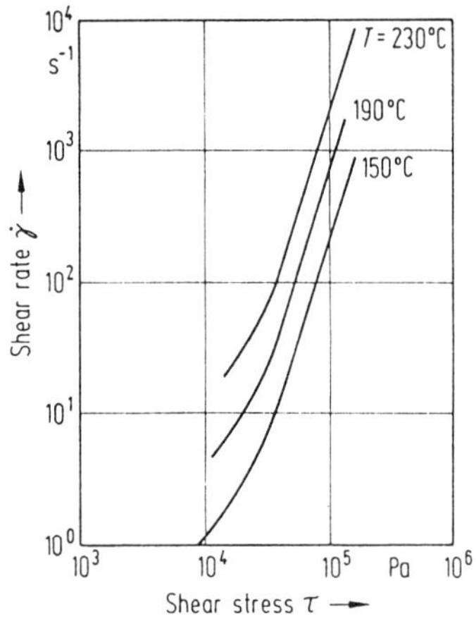

Fig. 3.7: Flow curves of a PE-LD [2]

3.2.4 Apparent Viscosity

The apparent viscosity η_a is defined as

$$\eta_a = \frac{\tau}{\dot{\gamma}_a} \tag{3.2.4}$$

and is shown in Fig. 3.8 as a function of shear rate and temperature for a PE-LD. Viscosity functions for several polymers are given in Figs. 3.9 and 3.10.

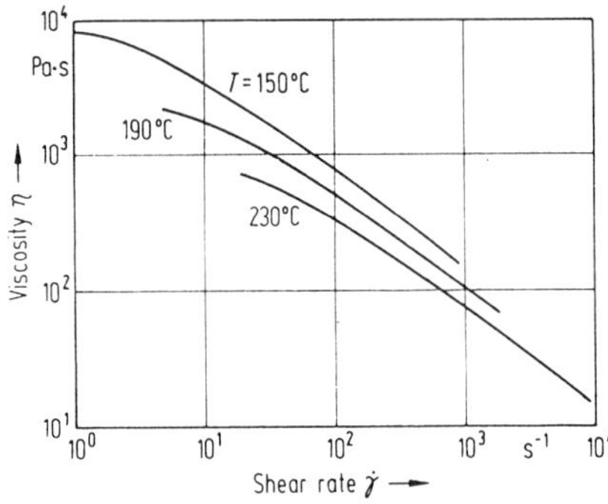

Fig. 3.8: Viscosity functions of a PE-LD [2]

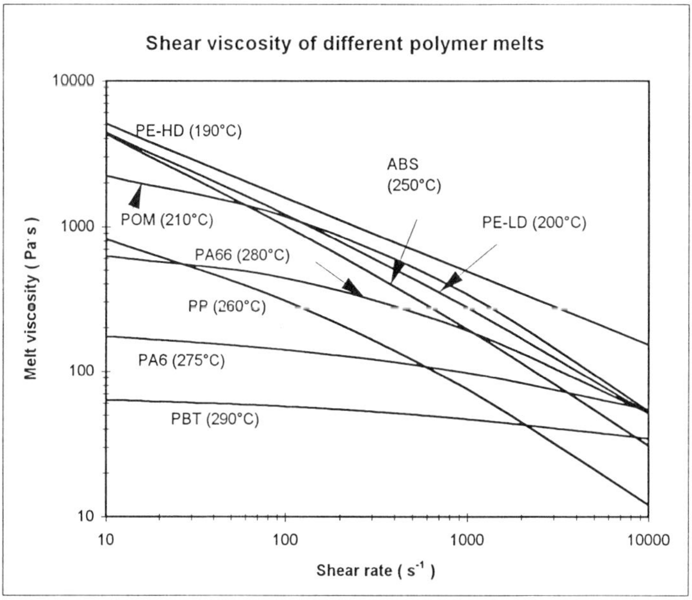

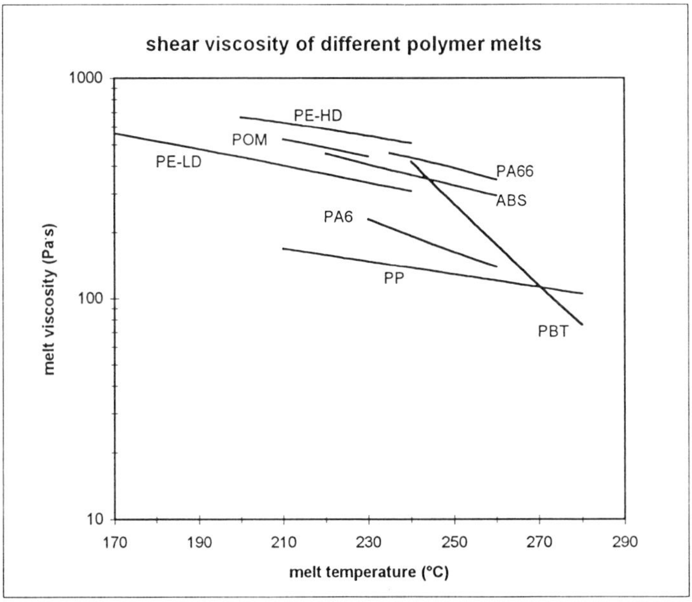

Fig. 3.9: Shear viscosity of some polymer melts

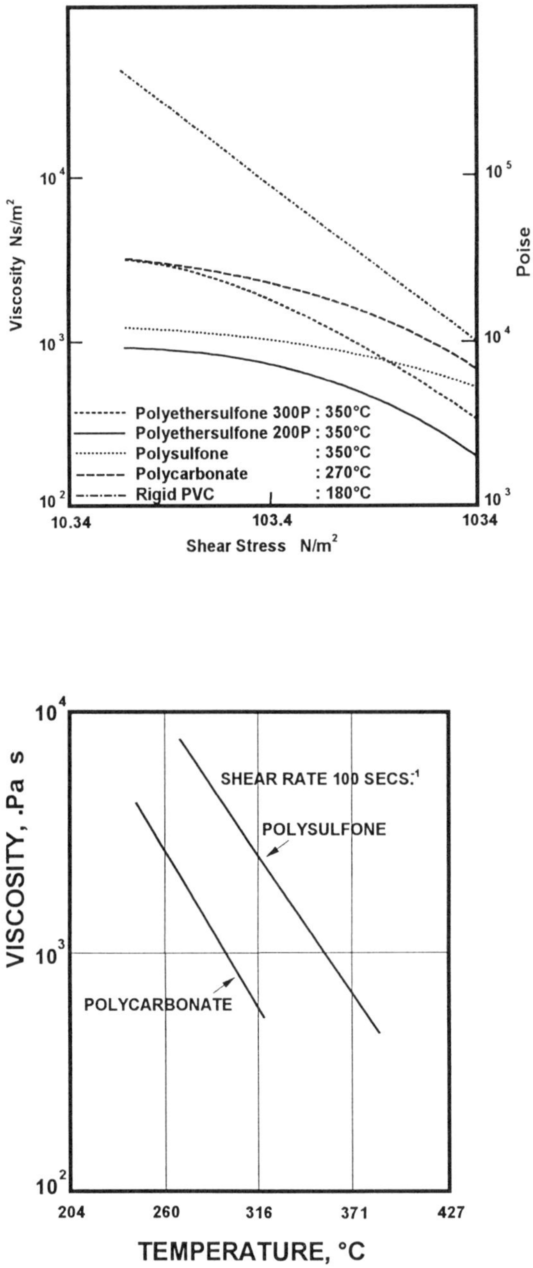

Fig. 3.10: Shear viscosity of some engineering thermoplastics [3], [16]

3.2.5 True Shear Rate

The true shear rate $\dot{\gamma}_t$ is obtained from the apparent shear rate by applying the correction for the non-Newtonian behavior of the melt according to Rabinowitsch

$$\dot{\gamma}_t = \left(\frac{n+3}{4}\right)\dot{\gamma}_a \tag{3.2.5}$$

The meaning of the power law exponent n is explained in Section 3.3.2.

3.2.6 True Viscosity

The true viscosity η_w is given by

$$\eta_w = \frac{\tau}{\dot{\gamma}_t} \tag{3.2.6}$$

In Fig. 3.11 the true and apparent viscosities are plotted as functions of the corresponding shear rates at different temperatures for a polystyrene. As can be seen, the apparent viscosity function is a good approximation for engineering calculations.

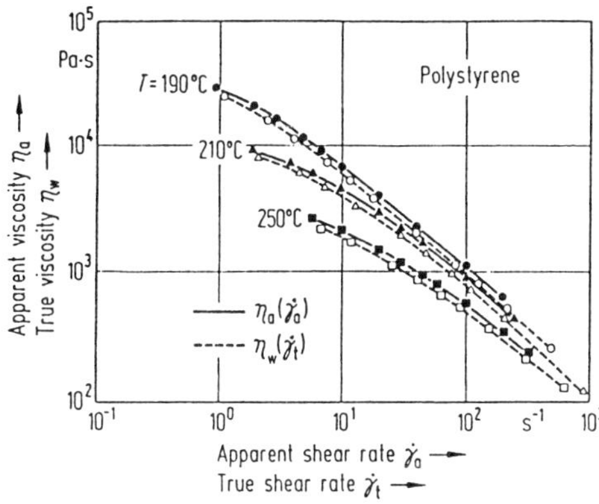

Fig. 3.11: True and apparent viscosity functions of a polystyrene at different temperatures [4]

3.3 Rheological Models

Various fluid models have been developed to calculate the apparent shear viscosity η_a [2]. The following sections deal with an important few of these relationships, which are frequently used in design calculations.

3.3.1 Hyperbolic Function of Prandtl and Eyring

The relation between shear rate $\dot{\gamma}_a$ and shear stress τ according to the fluid model of Eyring [19] and Prandtl [20] can be written as

$$\dot{\gamma}_a = -C \sinh(\tau / A) \tag{3.3.1}$$

where C and A are temperature-dependent material constants.

The evaluation of the constants C and A for the flow curve of PE-LD at 190°C in Fig. 3.12 leads to $C = 4 s^{-1}$ and $A = 3 \cdot 10^4$ N/m². It can be seen from Fig. 3.12 that the hyperbolic function of Prandtl and Eyring holds good at low shear rates.

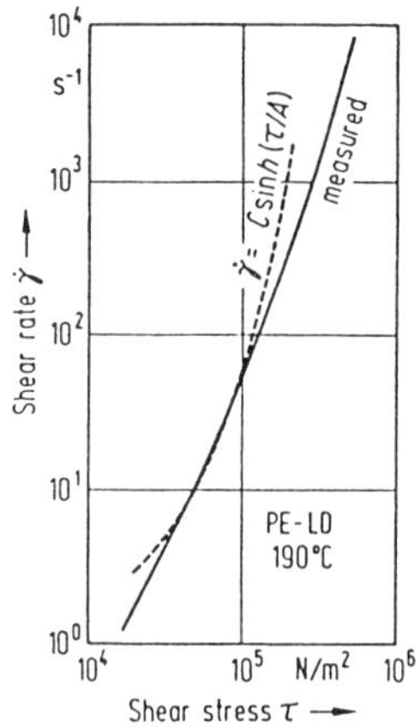

Fig. 3.12: Comparison between measurements and values calculated with Eq. (3.3.1) [2]

3.3.2 Power Law of Ostwald and De Waele

The power law of Ostwald [21] and De Waele [22] is easy to use, hence widely employed in design work [5]. This relation can be expressed as

$$\dot{\gamma}_a = K \tau^n \tag{3.3.2}$$

or

$$\dot{\gamma}_a = K \left| \tau^{n-1} \right| \tau \tag{3.3.3}$$

where K denotes a factor of proportionality and n the power law exponent. Another form of power law often used is

$$\tau_a = K_R \dot{\gamma}_a^{n_R} \tag{3.3.4}$$

or

$$\tau_a = K_R \left|\dot{\gamma}_a^{n_R - 1}\right| \dot{\gamma}_a \tag{3.3.5}$$

In this case, n_R is the reciprocal of n and $K_R = K^{-n_R}$.
From Eq. (3.3.2) the exponent n can be expressed as

$$n = \frac{d \lg \dot{\gamma}_a}{d \lg \tau} \tag{3.3.6}$$

As shown in Fig. 3.13 in a double log-plot the exponent n represents the local gradient of the curve $\dot{\gamma}_a$ vs. τ.

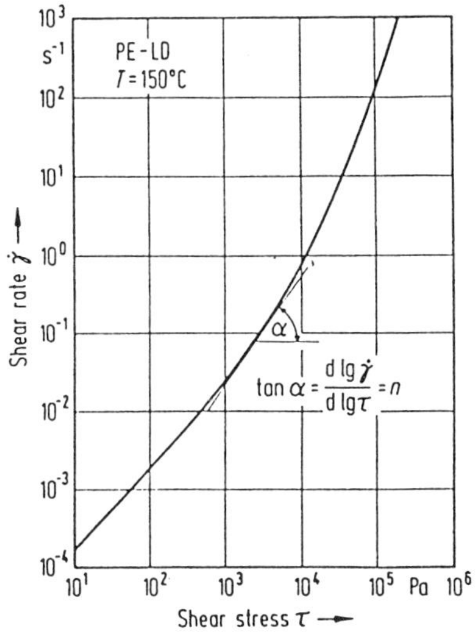

Fig. 3.13: Determination of the power law exponent n in the Eq. (3.3.2)

Furthermore

$$\frac{1}{n} = \frac{d \lg \tau}{d \lg \dot{\gamma}_a} = \frac{d \lg \eta_a + d \lg \dot{\gamma}_a}{d \lg \dot{\gamma}_a} = \frac{d \lg \eta_a}{d \lg \dot{\gamma}_a} + 1 \tag{3.3.7}$$

The values of K and n determined from the flow curve of PE-LD at 190°C shown in Fig. 3.14 were found to be $K = 1.06 \cdot 10^{-11}$ and $n = 2.57$.
As can be seen from Fig. 3.14, the power law fits the measured values much better than the hyperbolic function of Prandtl [20] and Eyring [19]. The deviation between the power law and experiment is a result of the assumption that the exponent n is constant throughout

the range of shear rates considered, whereas actually n varies with the shear rate. The power law can be extended to consider the effect of temperature on viscosity as follows:

$$\eta_a = K_{OR} \cdot \exp(-\beta \cdot T) \cdot \dot{\gamma}_a^{n_R - 1} \tag{3.3.8}$$

where K_{OR} = consistency index

β = temperature coefficient

T = temperature of melt.

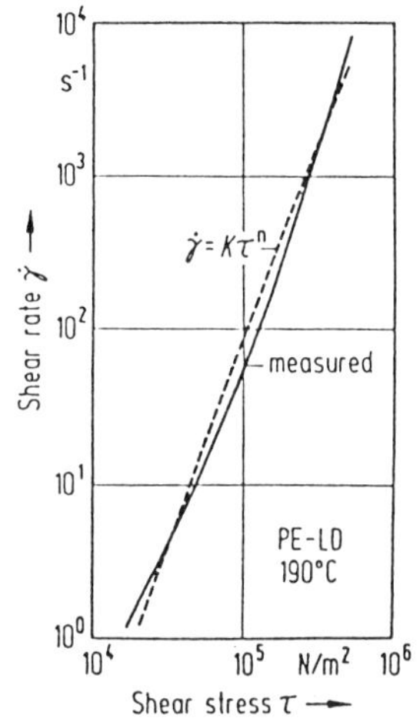

Fig. 3.14: Comparison between measured values and power law

Example:

Following values are given for a PE-LD:

n_R = 0.3286

β = 0.00863°C^{-1}

K_{OR} = $135990\ N \cdot s^{n_R} \cdot m^{-2}$

The viscosity η_a at T= 200°C and $\dot{\gamma}_a$=500 s^{-1} is calculated from Eq. (3.3.8)

η_a = 373.1 $Pa{\cdot}s$

3.3.3 Polynomial of Muenstedt

The fourth degree polynomial of Muenstedt provides a good fit for the measured values of viscosity. For a definite temperature this is expressed as

$$\lg \eta_a = A_0 + A_1 \lg \dot{\gamma}_a + A_2 (\lg \dot{\gamma}_a)^2 + A_3 (\lg \dot{\gamma}_a)^3 + A_4 (\lg \dot{\gamma}_a)^4 \tag{3.3.9}$$

where A_0, A_1, A_2, A_3, A_4 represent resin-dependent constants. These constants can be determined with the help of the program of Rao [13], which is based on multiple linear regression.

This program in its general form fits an equation of the type $y=a_0+a_1x_1+a_2x_2+...a_nx_n$ and prints out the coefficients a_0, a_1 and so on for the best fit.

3.3.3.1 Shift Factor for Crystalline Polymers

The influence of temperature on viscosity can be taken into account by the shift factor a_T [4]. For crystalline polymers this can be expressed as

$$a_T = b_1(T_0)\exp(b_2/T) \tag{3.3.10}$$

where b_1, b_2 = resin-dependent constants
T = melt temperature (K)
T_0 = reference temperature (K)

3.3.3.2 Shift Factor for Amorphous Polymers

The shift factor a_T for amorphous polymers is derived from the WLF equation and can be written as

$$\lg a_T = \frac{-c_1(T-T_0)}{c_2+(T-T_0)} \tag{3.3.11}$$

where c_1, c_2 = resin-dependent constants
T = melt temperature (°C)
T_0 = reference temperature (°C)

The expression for calculating both the effect of temperature and shear rate on viscosity follows from Eq. (3.3.9)

$$\begin{aligned}\lg \eta_a = {} & \lg a_T + A_0 + A_1 \lg(a_T\dot{\gamma}_a) + A_2[\lg(a_T\dot{\gamma}_a)]^2 \\ & + A_3[\lg(a_T\dot{\gamma}_a)]^3 + A_4[\lg(a_T\dot{\gamma}_a)]^4\end{aligned} \tag{3.3.12}$$

With Eq. (3.3.7) we get

$$\frac{1}{n} = 1 + A_1 + 2A_2\lg(a_T\dot{\gamma}_a) + 3A_3[\lg(a_T\dot{\gamma}_a)]^2 + 4A_4[\lg(a_T\dot{\gamma}_a)]^3 \tag{3.3.13}$$

The power law exponent is often required in the design work as a function of shear rate and temperature. Fig. 3.15 illustrates this relationship for a definite PE-LD. The curves shown are computed with Eqns. (3.3.10) and (3.3.13). As can be inferred from Fig. 3.15, the assumption of a constant value for the power law exponent holds good for a wide range of shear rates.

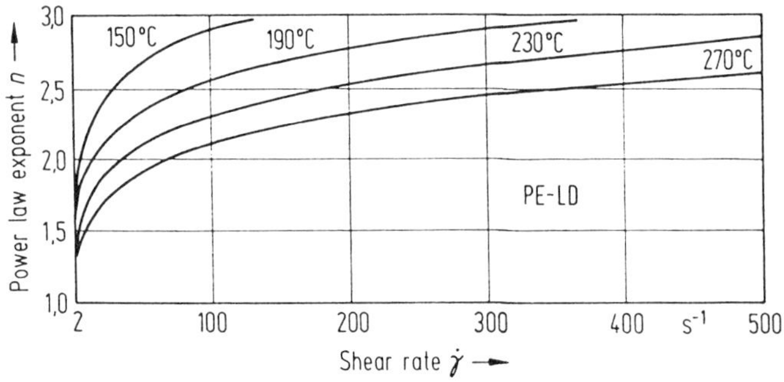

Fig. 3.15: Power law exponent of a PE-LD as a function of shear rate and temperature

Example:

A_0 = 4.2541
A_1 = -0.4978
A_2 = -0.0731
A_3 = 0.0133
A_4 = -0.0011
b_1 = $-5.13 \cdot 10^{-6}$
b_2 = $5640\ K$

at $\dot{\gamma}_a = 500\ s^{-1}$ and T= $200°C$.

Solution:

a_T from Eq. (3.3.10)

$a_T = 5.13 \cdot 10^{-6} \cdot \exp(5640/473) = 0.774$

With $X = \lg(a_T \cdot \dot{\gamma}_a)$

$X = \lg(0.774 \cdot 500) = 2.588$

η_a from Eq. (3.3.12)

$$\eta_a = 10^{(\lg a_T + A_0 + A_1 X + A_2 X^2 + A_3 X^3 + A_4 X^4)}$$

Substituting the values of A_0, A_1 and so on one gets

$$\eta_a = 351.78\ Pa \cdot s$$

The power law exponent is obtained from Eq. (3.3.13)

$$n = (1 + A_1 + 2A_2 X + 3A_3 X^2 + 4A_4 X^3)^{-1}$$

Using the values of A_0, A_1 and so on

$n = 3.196$

3.3.4 Viscosity Equation of Carreau [23]

As shown in Fig. 3.16, the Carreau equation gives the best fit for the viscosity function reproducing the asymptotic form of the plot at high and low shear rates correctly.

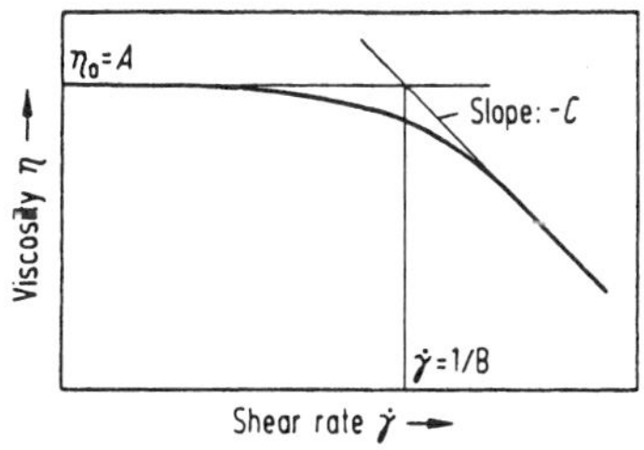

Fig. 3.16: Determination of Carreau parameters from a viscosity function [9]

The equation is expressed as

$$\eta_a = \frac{A}{(1 + B\ \dot{\gamma}_a)^C} \tag{3.3.14}$$

where A, B, C are resin-dependent constants. By introducing the shift factor a_T into Eq. (3.3.14) the temperature-invariant form of the Carreau equation can be given as (Fig. 3.17)

$$\eta_a = \frac{A a_T}{(1 + B a_T \dot{\gamma}_a)^C} \tag{3.3.15}$$

For a number of resins the shift factor can be calculated as a function of temperature from the following equation with good approximation [5], [6]

$$\lg a_T(T_1, T_2) = \frac{8.86(T_1 - T_{ST})}{101.6 + (T_1 - T_{ST})} - \frac{8.86(T_2 - T_{ST})}{101.6 + (T_2 - T_{ST})} \tag{3.3.16}$$

where T_1 (°C) is the temperature, at which the viscosity is given and T_2 (°C), the temperature, at which the viscosity is to be found out.

The standard temperature T_{ST} is given by [6]

$$T_{ST} = T_g + 50°\,C \tag{3.3.17}$$

Data on typical glass transition temperatures of polymers are given in Table 2.4.
The power law exponent n can be obtained from Eq. (3.3.15):

$$\frac{1}{n} = -C \cdot \frac{B \cdot a_T}{1 + B \cdot a_T \cdot \dot{\gamma}_a} \cdot \dot{\gamma}_a + 1 \tag{3.3.18}$$

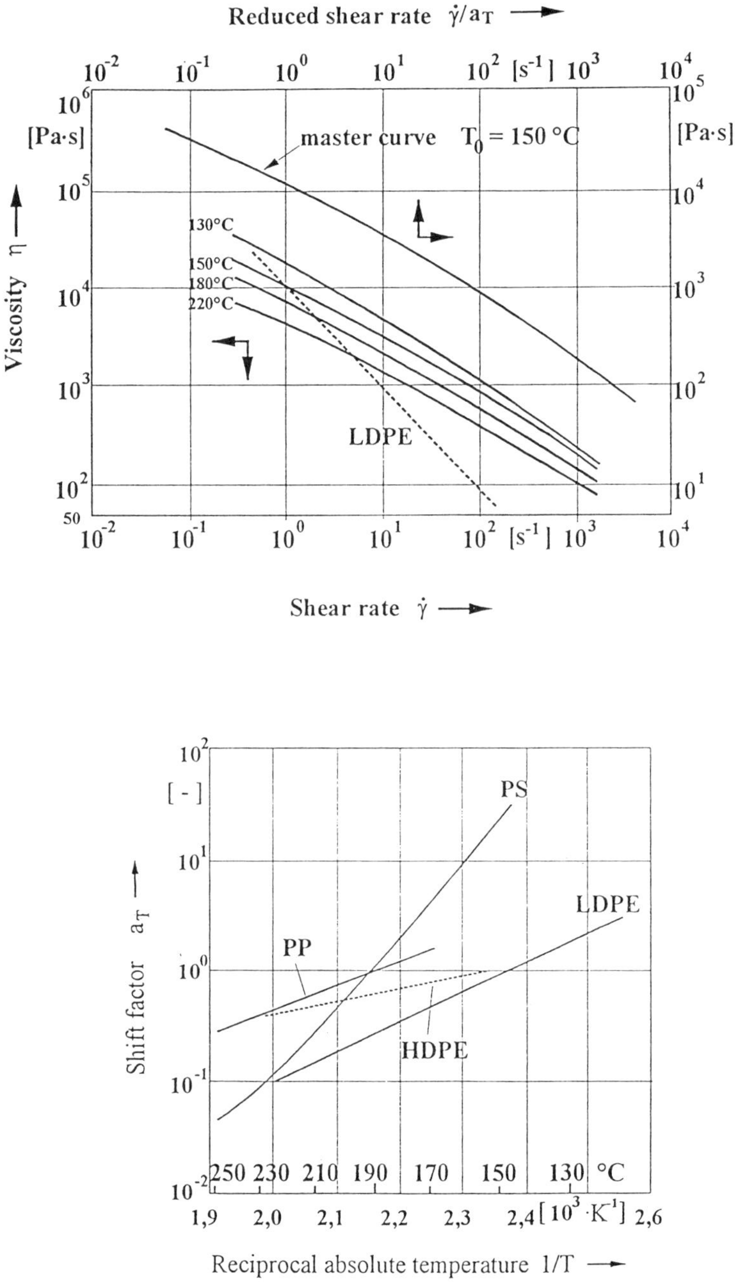

Fig. 3.17: Use of shift factor a_T for calculating temperature invariant viscosity [17]

For high shear rates n becomes [4]

$$n = \frac{1}{1-C} \tag{3.3.19}$$

Computer disks containing the resin-dependent constants *A*, *B*, *C* can be obtained for the respective resins from the resin manufacturers [12]. These constants can also be determined by using the software VISRHEO [13], and can be stored in a data bank for viscosity data.

Example:
Following constants are given for a particular PE-LD.

A	=	32400 Pa · s
B	=	3.1 s
C	=	0.62
T_{ST}	=	-133 C
T_1	=	190 C

The viscosity is to be calculated at

$T_2 = 200°C$ and $\dot{\gamma}_a = 500\ s^{-1}$

Solution:
One obtains from Eq. (3.3.16)

$$X = \frac{8.86(T_1 - T_{ST})}{101.6 + (T_1 - T_{ST})} = \frac{8.86(190-(-133))}{101.6+(190-(-133))} = 6.74$$

and

$$Y = \frac{8.86(T_2 - T_{ST})}{101.6 + (T_2 - T_{ST})} = \frac{8.86(200-(-133))}{101.6+(200-(-133))} = 6.79$$

The power law exponent is calculated from Eq. (3.3.18)

$$n = \left(\frac{-C \cdot Z}{1+Z} + 1\right)^{-1} = \left(\frac{-0.62 \cdot 1379.5}{1+1379.5} + 1\right) = 2.63$$

The viscosity η_a follows from Eq. (3.3.15)

$$\frac{32400 \cdot 0.89}{(1+1379 \cdot 5) \cdot 0.62} = 325.8 \text{ Pa s}$$

3.3.5 Viscosity Formula of Klein

The regression equation of Klein et al. [25] is given by

$$\lg \eta_a = a_0 + a_1 \ln \dot{\gamma}_a + a_{11}(\ln \dot{\gamma}_a)^2 + a_2 T + a_{22} T^2 + a_{12} T \ln \dot{\gamma}_a \tag{3.3.20}$$

T = Temperature of the melt (°*F*)

η_a = Viscosity ($lb_f s/in^2$)

The resin-dependent constants a_0 to a_{22} can be determined with the help of the computer program given in [13], as has been the case in finding out the A-coefficients in Eq. (3.3.9).

Example:
Following constants are valid for a particular type of PE-LD. What is the viscosity η_a at $\dot{\gamma}_a$=500 s^{-1} and T= 200°C ?

a_0 = 3.388
a_1 = $-6.351 \cdot 10^{-1}$
a_{11} = $-1.815 \cdot 10^{-2}$
a_2 = $-5.975 \cdot 10^{-3}$
a_{22} = $-2.51 \cdot 10^{-6}$
a_{12} = $5.187 \cdot 10^{-4}$

Solution:

$$T(°F) = 1.8 \cdot T(°C) + 32 = 1.8 \cdot 200 + 32 = 392$$

With the constants above and Eq. (3.3.20) one gets

$$\eta_a = 0.066\ lb_f{\cdot}s/in^2$$

and in SI-units

$$\eta_a = 6857 \cdot 0.066 = 449.8\ Pa \cdot s$$

The expression for the power law exponent n can be derived from Eq. (3.3.7) and Eq. (3.3.19). The exponent n is given by

$$\frac{1}{n} = 1 + a_1 + 2a_{11} \ln \dot{\gamma}_a + a_{12} \cdot T \qquad (3.3.21)$$

Putting the constants a_1, ..., a_{12} into this equation one obtains

$$n = 2.919$$

3.4 Effect of Pressure on Viscosity

Compared to the influence of temperature the effect of pressure on viscosity is not of much significance.
However, the relative measure of viscosity can be obtained from [8], [6], [4]

$$\eta_p = \eta_0 \exp(\alpha_p \cdot p) \qquad (3.4.1)$$

where η_p = viscosity at pressure p and constant shear stress τ_0
η_0 = viscosity at constant shear stress τ_0
α_p = pressure coefficient

For styrene polymers η_p is calculated from [8]

$$\eta_p = \eta_0 \exp(p / 1000) \qquad (3.4.2)$$

where p = pressure in bar.

Thus the change of viscosity with the pressure can be obtained from Eq. (3.4.2). Table 3.2 shows the values of viscosity calculated according to Eq. (3.4.2) for a polystyrene of average molecular weight. It can be seen that a pressure of 200 bar causes an increase of viscosity of 22% compared to the value at 1 bar. The pressure coefficient of PE-LD is less than that of PS by factor of 3 to 4 and the value PE-HD is again less than a factor of 2 than that of PE-LD. This means that in the case of polyethylene an increase of pressure by 200 bar would enhance the viscosity only by 3 to 4%. Consequently, the effect of pressure on viscosity can be neglected in the case of extrusion processes, in which generally low pressures exist. However, in injection molding where usually one has to deal with high pressures the dependence of viscosity on pressure has to be considered.

Table 3.2: Effect of Pressure on Viscosity for Polystyrene, Eq. (3.4.2)

p bar	η_p
30	$1.03\,\eta_0$
100	$1.105\,\eta_0$
200	$1.221\,\eta_0$
300	$1.35\,\eta_0$
500	$1.65\,\eta_0$
1000	$2.72\,\eta_0$
3000	$20\,\eta_0$

Figure 3.18 shows the melt viscosity at constant stress and temperature as a function of pressure for some polymers [7].

3.5 Dependence of Viscosity on Molecular Weight

The relationship between viscosity and molecular weight can be described by [10]

$$\eta_a = K' M_w^{3.5} \tag{3.5.1}$$

where M_w = molecular weight

K' = resin dependent constant.

The approximate value of K' for PE-LD is

$$K' = 2.28 \cdot 10^{-4}$$

and for Polyamide 6

$$K' = 5.21 \cdot 10^{-14}$$

according to the measurements of Laun [10]. These values are based on zero viscosity.

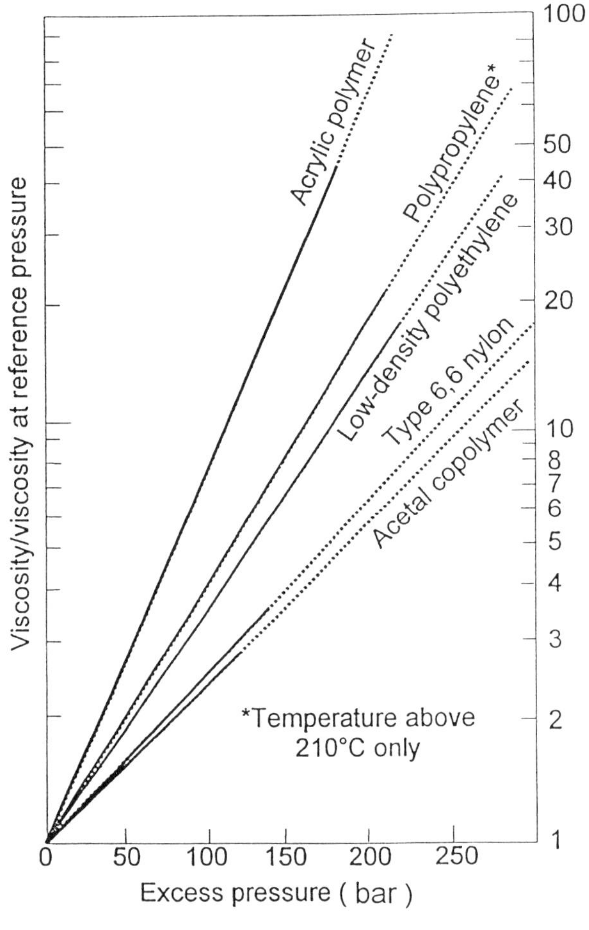

Fig. 3.18: Melt viscosity at constant stress and temperature as a function of pressure [7]

3.6 Viscosity of Two-Component Mixtures

The viscosity of a mixture consisting of the component *A* and the component *B* can be obtained from [11]

$$\lg \eta_M = C_A \lg \eta_A + C_B \lg \eta_B \tag{3.6.1}$$

where η = viscosity
C = weight percent

Indices:

M: mixture
A, B: components

3.7 Melt Flow Index

The Melt Flow Index (MFI) which is also known as the Melt Flow Rate (MFR) indicates the flowability of a constant polymer melt, and is measured by forcing the melt through a capillary under a dead load at constant temperature (Fig. 3.19). The MFI value is the mass of melt flowing in a certain time. A MFR or MFI of 2 at 200°C and 2.16 kg means, for example, that the melt at 200 °C flows at a rate of 2 g in ten minutes under a dead load of 2.16 kg.

In the case of Melt Volume Rate which is also known as Melt Volume Index (MVI) the volume flow rate of the melt instead of mass flow rate is set as the basis. The unit here is ml/10 min.

The effect of MFI on the properties of polyethylene, as an example, is illustrated in Fig. 3.20 [18]. Ranges of melt indices for common processing operations are given in Table 3.3 [18].

Table 3.3: Ranges of MFI Values (ASTM D1238) for Common Processes [18]

Process	MFI range
Injection molding	5/100
Rotational molding	5/20
Film extrusion	0.5/6
Blow molding	0.1/1
Profile extrusion	0.1/1

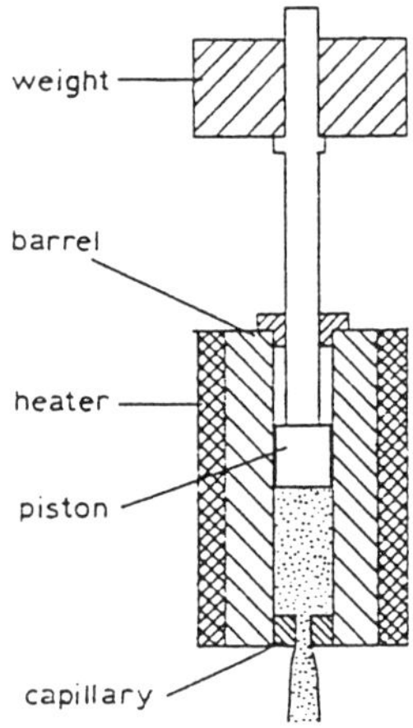

Fig. 3.19: Melt flow tester [6]

3.8 Tensile Viscosity

Although the flow of melt in the channels of dies and machines of polymer machinery is mainly shear flow, elongational flow is of importance in such applications as film blowing and blow molding. The elongational or tensile viscosity can be measured with a tensile rheometer [1], and is much higher than shear viscosity. As shown in Fig. 3.4 the tensile viscosity of a Newtonian fluid is three times the shear viscosity. The tensile viscosity is defined as

$$\mu = \frac{\sigma}{\dot{\varepsilon}} \tag{3.8.1}$$

$$\dot{\varepsilon} = \frac{1}{l} \cdot \frac{dl}{dt} \tag{3.8.2}$$

where l = length at any instant of a volume element
t = time

3.9 Viscoelastic Properties

Polymer machinery can be designed sufficiently accurate on the basis of the relationships for viscous shear flow alone. However, a complete analysis of melt flow should include both viscous and elastic effects, although the design of machines and dies by considering melt elasticity is rather difficult and seldom in use. Similar attempts to dimension the dies taking elastic effects into account have been made as described in the work of Wagner [26] and Fischer [27].

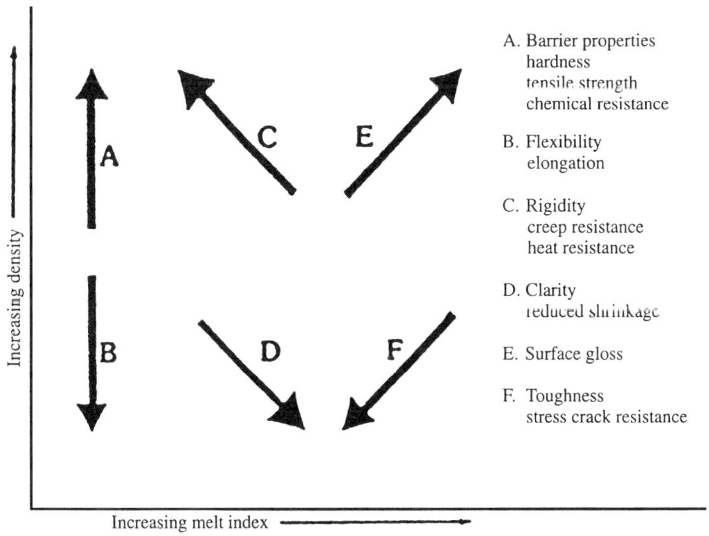

Fig. 3.20: Melt index and density vs. polymer properties [18]

To give a more complete picture of melt rheology the following expressions for the viscoelastic quantities according to Laun [10], [14] are presented.

The material functions characterizing the elastic behavior of a polymer melt are shear compliance and primary normal stress coefficient which are defined as follows [14], [2]:

3.9.1 Primary Normal Stress Coefficient Θ:

$$\Theta = \frac{N_1}{\dot{\gamma}_0^2} \tag{3.9.1}$$

N_l: normal stress difference, $\dot{\gamma}_0$: shear rate

3.9.2 Shear Compliance J_e:

$$J_e = \frac{\gamma_{r,s}}{\tau_0} \tag{3.9.2}$$

$\gamma_{r,s}$: recoverable shear strain, τ_o: shear stress

Further on we have [14]

$$J_e = \frac{\Theta}{2\eta^2} \tag{3.9.3}$$

The equations above shown as functions of shear rate can be determined from measurements with a cone and plate rheometer (Fig. 3.21) [1].

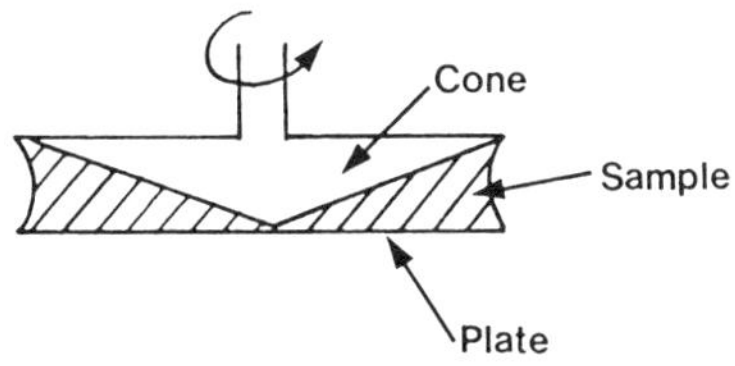

Fig. 3.21: Schematic diagram of a cone and plate rheometer [1]

The limiting values of these equations are Θ_0, η_0 and J_e^o (Fig. 3.22).

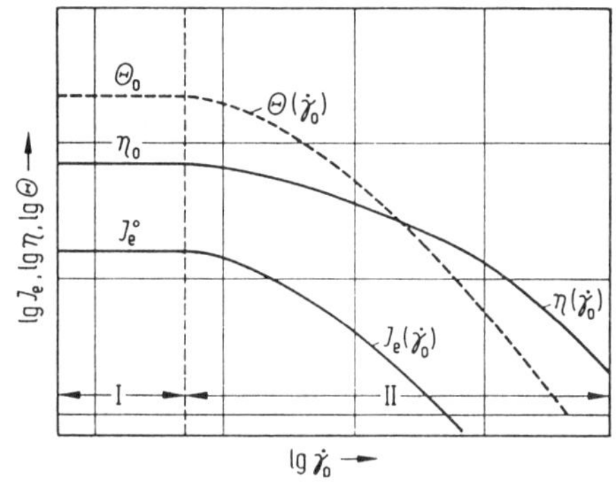

Fig. 3.22: Parameters for steady shear flow [14] (I = linear region, II = nonlinear region)

3.9.3 Die Swell

Die swell which can be measured with a capillary viscometer gives a measure of the elastic deformation of the melt. Die swell is shown in Fig. 3.23 [14] as a function of length L to radius R of the capillary. The value is highest for an orifice of negligible length where the effect of converging entrance flow is largest. With increasing L/R ratios the molecular orientation decays, and the swell attains a constant value. For certain applications smaller L/R ratios of dies are preferred in order to have a high molecular orientation.

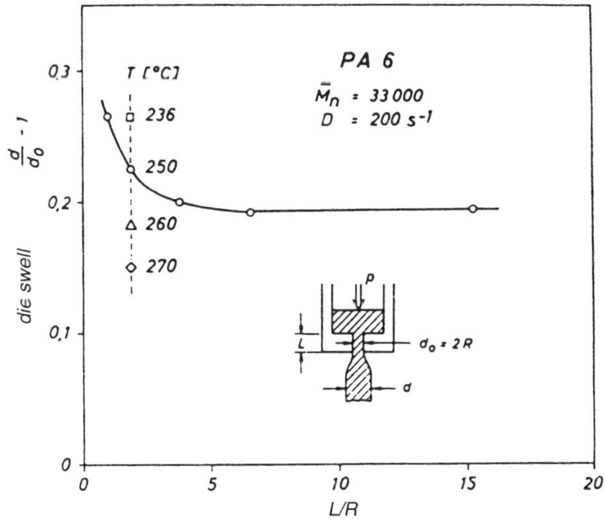

Fig. 3.23 Die swell vs length to radius L/R [14]

The viscoelastic behavior of polymer melts is treated in [2] in more detail.

Literature

1. *Birley, A.W.: Haworth, B.: Bachelor, T.:* Physics of Plastics, Hanser, Munich 1991
2. *Rao, N.S.:* Design Formulas for Plastics Engineers, Hanser, Munich, 1991
3. *Rigby, R.B.:* Polyethersulfone in Engineering Thermoplastics: Properties and Applications. Ed.: James M. Margolis. Marcel Dekker, Basel 1985
4. *Münstedt, H.:* Berechnen von Extrudierwerkzeugen, VDI-Verlag, Düsseldorf, 1978
5. *Rao, N.S.:* Designing Machines and Dies Polymer Processing, Hanser, Munich 1981
6. *Rauwendal, C.:* Polymer Extrusion, Hanser, Munich 1986
7. *Ogorkiewicz, R.M.:* Thermoplastics Properties and Design, John Wuley, New York 1973
8. *Avenas, P., Agassant, J.F., Sergent, J.Ph.:* La Mise en Forme des Materieres Plastiques, Technique & Documentation (Lavoirier), Paris 1982
9. *Hertlein, T., Fritz, H.G.:* Kunststoffe 78, 606 (1988)
10. *Laun, H.M.:* Progr. Colloid & Polymer Sai, 75, 111 (1987)
11. *Carley, J.F.:* ANTEC 84, p. 439
12. *CAMPUS Databank*
13. *Rao, N.S., O'Brien, K.T. and Harry, D.H.:* Computer Modeling for Extrusion and other Continous Polymer Processes. Ed. Keith T. O'Brien, Hanser, Munich 1992
14. *Laun, H.M.:* Rheol. Acta 18, 478 (1979)

15. *BASF Brochure:* Kunststoff-Physik im Gespräch, 1977
16. *Harris, J.E.:* Polysulfone in Engineering Thermoplastics: Properties and Applications. Ed.: James M. Margolis, Marcel Dekker, Basel 1985
17. *Geiger, K.:* Private Communication.
18. *Rosato, D.V., Rosato, D.V.:* Plastics Processing Data Handbook, Van Nostrand Reinhold, New York 1990
19. *Eyring, H.:* I. Chem. Phys. 4, 283 (1963)
20. *Prandtl, L.:* Phys. Blätter 5, 161 (1949)
21. *Ostwald, W.:* Kolloid-Z., 36, 99 (1925)
22. *De Waele, A.:* Oil and Color Chem. Assoc. J., 6, 33 (1923)
23. *Carreau, P.J.:* Dissertation, Univ. Wisconsin, Madison (1968)
24. *Bernhardt, E.C.:* Processing of Thermoplastic Materials, Reinbold, New York (1963)
25. *Klein, I.; Marshall, D.I.; Friehe, C.A.:* Soc Plastic Engrs. J. 21, 1299 (1965)
26. *Wagner, M.H.:* Dissertation, Univ. Stuttgart (1976)
27. *Fischer, E.:* Dissertation, Univ. Stuttgart (1983)

4 Electrical Properties

The use of plastics in the electrical industry as insulators for wire and cable insulation is well known. The application of engineering resins to make miniature electric components, printed circuit boards, conductive housings of computer equipment and the like, although not so well known, is increasing. Some of the electrical properties which are of importance in selecting a resin for these applications are treated in this section.

4.1 Surface Resistivity

Surface resistivity is defined as the ratio of electrical field strength to the current density in a surface layer of an insulating material. It is the measure of a material's ability to resist the flow of current along its surface when a direct voltage is applied between surface mounted electrodes of unit width and unit spacing. The unit of surface resistivity is ohm [1], [3].

4.2 Volume Resistivity

Volume resistivity is the volume resistance reduced to a cubical unit volume of the material. Volume resistance is the ratio of the direct voltage applied to the electrodes in contact with the test material, to the steady-state current flowing between them [1], [3]. The unit of volume resistivity is ohm·m (W·m) or ohm·cm (W·cm).

4.3 Dielectric Strength

Dielectric strength is the measure of the electrical breakdown resistance of a material under an applied voltage [1]. It is the ratio of the voltage reached just before breakdown to the material's thickness and is expressed as kv/mm or Mv/m. Dielectric strength data for some materials are given in Table 4.1.

Table 4.1: Dielectric Strength Data for Plastics (Method ASTM D149) [6]

Material	Dielectric strength kv/25μm
PE-LD	> 700
PE-HD	> 700
PP	800
PVC	300
PS	500
ABS	400
PC	350
POM	700
PA 6	350
PA 66	400
PMMA	300
PET	500
PBT	500

4.4 Relative Permittivity

Relative permittivity (ε_R), formerly known as dielectric constant is the ratio of capacitance (C) of a given configuration of electrodes with the plastics material as the dielectric medium, to the capacitance (C_v) of the same configuration of electrodes with vacuum as the dielectric [1], [3].

$$\varepsilon_R = \frac{C}{C_v} \tag{4.4.1}$$

The performance of plastics as insulators increases with decreasing relative permittivity.

4.5 Dielectric Dissipation Factor or Loss Tangent

Dielectric dissipation factor is the ratio of the electrical power dissipated in a material to the total power circulating in the circuit. It is the tangent of the loss angle (δ) and is analogous to tanδ which is the ratio between loss and storage moduli described in Section 1.3.2. A low dissipation factor is important for plastics insulators in high frequency applications such as radar and microwave equipment. Relative permittivity and dissipation factor are dependent on temperature, moisture, frequency and voltage [3]. Typical values of volume resistivity, relative permittivity and loss tangent are given in Table 4.2 for some polymers [1], [5] (see also Fig. 4.1 [2]).

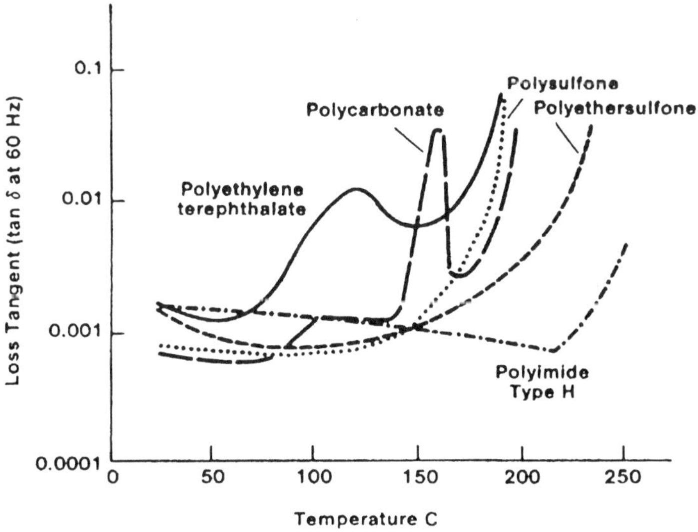

Fig. 4.1: Loss tangent as a function of temperature for some engineering thermoplastics [2]

Table 4.2: Typical Values of Volume Resistivity, Relative Permittivity and Loss Tangent [1], [5]

Material	Volume resistivity ohm × cm	Relative permittivity at 50 Hz	Dielectric dissipation factor or loss tangent $\tan\delta \cdot 10^4$ at 20°C and 1 MHz
PE-LD	10^{17}	2.3	1.2
PE-HD	10^{17}	2.35	2
PP	10^{17}	3.5	400
PVC	10^{15}	2.27	230
PS	10^{17}	2.5	1
ABS	10^{15}	2.9	200
PC	10^{17}	3	7
POM	10^{15}	3.7	50
PA 6	10^{12}	3.8	300
PA 66	10^{12}	8.0	800
PMMA	10^{15}	3.3	40
PET	10^{16}	4.0	200
PBT	10^{16}	3.0	200

4.6 Comparative Tracking Index (CTI)

Comparative tracking index indicates a plastics material's ability to resist development of an electrical conducting path when subjected to current in the presence of a contaminating solution [1], [3]. Contaminants, such as salt and moisture, allow increased conduction over the surface which may lead to tracking, the appearance of conducting paths over the surface. The deterioration of the surface quality is hence the cause of failure of high voltage insulation systems [1]. The CTI is given in terms of maximum voltage, at which no failure occurs, according to different test procedures. Typical values of CTI are given in Table 4.3 [5].

Table 4.3: Typical Values of Comparative Tracking Index [5]

Material	Test method KB
PE-LD	> 600
PE-HD	> 600
PP	> 600
PVC	600
PS	200
ABS	300
PC	> 600
POM	> 600
PA 6	> 600
PA 66	> 600
PMMA	> 600
PBT	450

Owing to their high electrical resistance plastics retain electrostatic charge which leads to such undesirable effects as marked attraction of dust or creation of discharges when the material comes into contact with other surfaces. This situation can be alleviated by using anti-static agents in the polymer formulation. Polymers with intrinsic electrical conductivity can also be used as anti-static coatings [1].

In addition to the properties treated in the foregoing sections the optical properties of plastics products such as transparency and gloss of films play an important role in selecting a resin in applications where these properties are required. The test procedures to measure the optical properties are treated in [1].

A list of properties for which data can be obtained from the resin manufacturers, for example, on CAMPUS computer disks is presented in Table 4.4. Example of material data used as input to the software VISMELT [4] for designing extruders is given in Table 4.5.

Table 4.4: List of Properties Obtainable from Resin Manufacturers

Mechanical properties

Tensile strength
Tensile elongation
Tensile modulus
Flexural strength
Flexural modulus
Compressive strength
Hardness
Abrasion
Izod impact strength

Thermal properties

Material density
Bulk density
pvT diagrams
Specific heat
Enthalpy
Thermal conductivity
Vicat temperature
HDT (Heat distortion temperature)
Flammability UL 94 rating
LOI (Limiting Oxygen Index)

Electrical properties

Dielectric strength
Surface resistivity
Volume resistivity
Relative permittivity
Dielectric dissipation factor or loss tangent
Comparative Tracking Index (CTI)

Optical properties

Light transmission
Refractive Index

Rheological properties

Melt Flow Rate (MFR)
Melt Volume Rate (MVR)
Shear viscosity as a function of shear rate and temperature of melt

Table 4.5: Example of Input Data to the Software Design Package VISMELT [4]

```
Type of Polymer:         PE-HD
Trade name of polymer:   HOECHST GM 9255 F

thermal properties :

melting point                       TM     =  130.0    Grad Celsius
specific heat of melt               CPM    =    2.51   kJ / (kg K)
specific heat of solid              CPS    =    2.30   kJ / (kg K)
thermal conductivity of melt        KM     =     .2700 W / (m K)
thermal conductivity of solid       KS     =     .2800 W / (m K)
heat of fusion                      LAM    =  200.000  kJ / kg
densitiy of melt                    RHOM   =     .78   g / cm**3
density of solid                    RHOS   =     .9430 g / cm**3
bulk density                        RHOS0  =     .40   g / cm**3

Viscosity coefficients :

Carreau-coefficients :
A     =  28625. Pa s                  B     =      .7126 s
C     =       .6535
T0    =    200.0 Grad Celsius
b     =   2447.  K

Muenstedt-coefficients :
A0    =      4.2410                   A1    =     -.01304
A2    =      -.55251                  A3    =      .222252
A4    =      -.033758
T0    =    200.0 Grad Celsius
b     =   2425.  K

Klein-coefficients :
```

```
A0   =   .653721E+01                         A1   =  -.722213E+00
A11  =  -.989580E-02                         A2   =  -.213261E-01
A22  =   .204288E-04                         A12  =   .473881E-03

Power law coefficients :
NR   =        .4022
BETA =        .003919 1/Grad Celsius     KOR  =  60889. N s**n / m**2
```

Literature

1. *Birley, A.W.: Haworth, B.: Bachelor, T.:* Physics of Plastics, Hanser, Munich 1991
2. *Rigby, R.B.:* Polyethersulfone in Engineering Thermoplastics: Properties and Applications. Ed.: James M. Margolis. Marcel Dekker, Basel 1985
3. *General Electric Plastics Brochure:* Engineering Materials Design Guide
4. *Rao, N.S., O'Brien, K.T. and Harry, D.H.:* Computer Modeling for Extrusion and other Continous Polymer Processes. Ed. Keith T. O'Brien, Hanser, Munich 1992
5. *BASF Brochure:* Kunststoff-Physik im Gespräch, 1977
6. *Domininghaus, H.:* Plastics for Engineers, Hanser, Munich 1993

5 Optical Properties of Solid Polymers

5.1 Light Transmission

The intensity of light incident on the surface of a plastic is reduced as the light enters the plastic because some light is always reflected away from the surface. The intensity of light entering the plastic is further reduced as the light passes through the plastic since some light is absorbed, or scattered, by the plastic. The luminous transmittance is defined as the percentage of incident light which is transmitted through the plastic. For comparison purposes the exact test parameters are documented in ASTM D 1003. Some typical light transmission values are presented in Table 5.1 for most common optical plastics. Light transmission is a measurement of the transparency of a plastic.

5.2 Haze

Haze is defined as the percentage of transmitted light which deviates from the incident light beam by more than 2.5 degrees. Its measurement is also defined by ASTM D 1003. Some typical haze values are presented in Table 5.2 for most common optical plastics. Haze is a measure of the clarity of a plastic.

5.3 Refractive Index

The refractive index n of an isotropic material is defined as the ratio of the speed of light in the material v to the speed of light in a vacuum c, that is,

$$n = v/c$$

The speed of light in vacuum is 300 Mm/s. The refractive index decreases as the wavelength of the light increases. Therefore the refractive index is measured and reported at a number of standard wavelengths, or atomic emission spectra (AES) lines, as indicated in Table 5.3.

Table 5.1: Light Transmission, or Luminous Transmittance of Some Common Optical Plastics

Material	Luminous Transmittance D 1003
ABS	85
PC	89
PMMA	92
PMMA/PS	90
PS	88
SAN	88

Table 5.2: Haze of Some Common Optical Plastics

Material	Haze
ABS	10
PC	1–3
PMMA	1–8
PMMA/PS	2
PS	3
SAN	3

Table 5.3: Refractive Indices as Functions of Wavelength

AES line	Wavelength	PMMA	PS	PC
F	486 nm	1.497	1.607	1.593
D	589 nm	1.491	1.590	1.586
C	651 nm	1.489	1.584	1.576

The refractive index is usually measured using an Abbe refractometer according to ASTM D542. The Abbe refractometer also measures the dispersions, which is required for lens design. An extensive list of refractive indices is provided in Table 5.4. Since speed of light in the polymer v is a function of the density, polymers which exhibit a range of densities also exhibit a range of refractive indices. Since density is a function of crystallinity, the refractive index is dependent on whether the polymer is amorphous or crystalline, and on its degree of crystallinity. Since density is also a function of temperature, decreasing as temperature increases, the refractive index also decreases with increasing temperature.

Table 5.4: Refractive Indices of Some Plastics as Functions of Density [1]

Polymer	Refractive index n_D^{20}	Density kg/m^3	Transparency
PE-LD	1.51	914/928	transparent
PE-HD	1.53	940/960	opaque
PP	1.50/1.51	890/910	transparent/opaque
PVC	1.52/1.55	1380/1550	transparent/opaque
PS	1.59	1050	transparent
PC	1.58	1200	transparent
POM	1.48	1410/1420	opaque
PA 6	1.52	1130	transparent/opaque
PA 66	1.53	1140	transparent/opaque
PMMA	1.49	1170/1200	transparent
PET	1.64	1380	transparent/opaque

5.4 Gloss

Surface gloss is the percentage of light intensity reflected relative to that reflected by an ideal surface. Also called specular reflectance or specular gloss, the gloss is measured within a specific angular range of the ideal reflected angle. It is then compared to a standard, polished black glass with a refractive index of 1.567, which possesses a specular gloss of 100%. The gloss reduces rapidly as the surface roughness increases. The specular gloss is measured according to ASTM D 523-80.

5.5 Color

Color may be measured using tristimulus colorimeters (a.k.a. colorimeters), or spectrophotometers (a.k.a. spectrocolorimeters). These instruments measure color by illuminating a sample and collecting the reflected light. Colorimeters use filters to simulate the color response of the eye. They are used for quick and simple measurements of color differences. Spectrocolorimeters measure color as a certain wavelength of light. The color is reported as three numbers. The L-value measures the grayness, with pure white scoring 100 and pure black scoring 0. The a-value measures the redness when positive and the greenness when negative. The b-value measures the yellowness when positive and the blueness when negative. So from the L, a, and b values a color can be defined.

Literature

1. *Domininghaus, H.:* Plastics for Engineers, Hanser, Munich 1993

6 External Influences

Plastics parts are often used in an environment, in which an interaction between the polymer and a fluid like water, gas or organic liquids can take place. This may lead to deterioration or even failure of the part. Furthermore, plastics used in certain applications as in packaging foodstuffs should be resistant to the entry of oxygen or moisture, so that the contents do not deteriorate during the prescribed time span.

The behavior of polymers under external influences is dependent on the combination of polymer and fluid, exposure time, temperature, stress level and processing history [1]. The properties of polymers which manifest themselves under external effects are the topic of this section.

6.1 Physical Interactions

6.1.1 Solubility

Most thermoplastics have a solubility parameter [1], [3]. Common solvents are paraffins, ethers, ketones, alcohols and chlorinated organic liquids. The solvation leads to swelling, weight and dimensional changes together with property loss.

6.1.2 Environmental Stress Cracking (ESC)

ESC is a failure mechanism which occurs under conditions when an external or residual stress is imposed on the part which is in contact with an external environment such as liquid or vapor. It is the combination of stress and the liquid medium, which gives rise to premature failure. The failure is initiated by microcrazing into which the aggressive liquid or vapor penetrates [1]. The term ESC is applied to amorphous polymers, where as in the case of crystalline polymers the failure is denoted as stress corrosion failure.

One of the standard ESC tests is the Bell Telephone technique, in which bent strips of material containing a defect are totally immersed in a chemical medium before being examined for visible signs of damage (Fig. 6.1) [1]. The environmental behavior of some plastics is presented in Fig. 6.2 [7] and Fig. 6.3 [4].

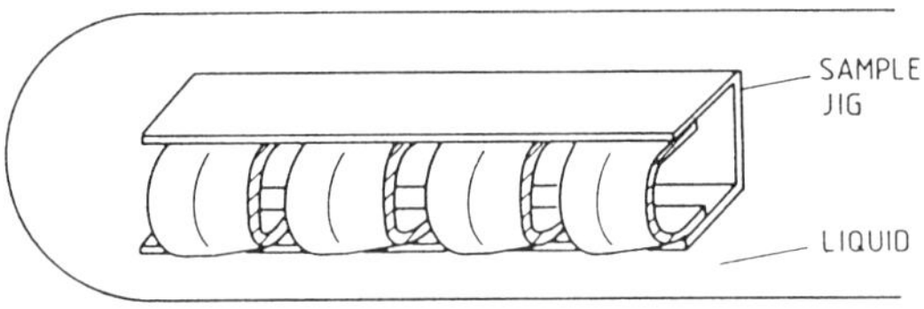

Fig. 6.1: Bell test for determining ESC [1]

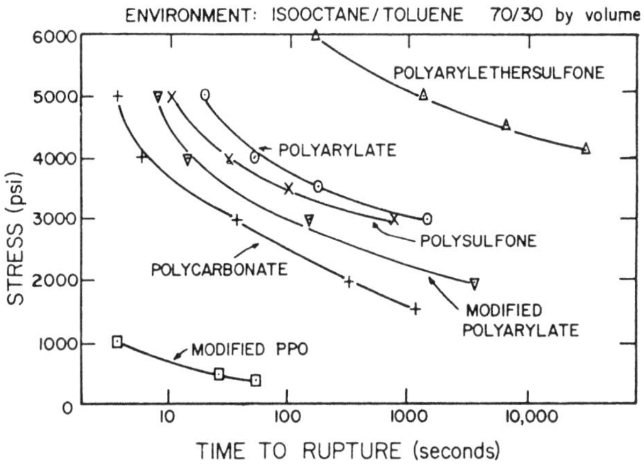

Fig. 6.2: ESC resistance of polyarylate vs. other thermoplastics [7]

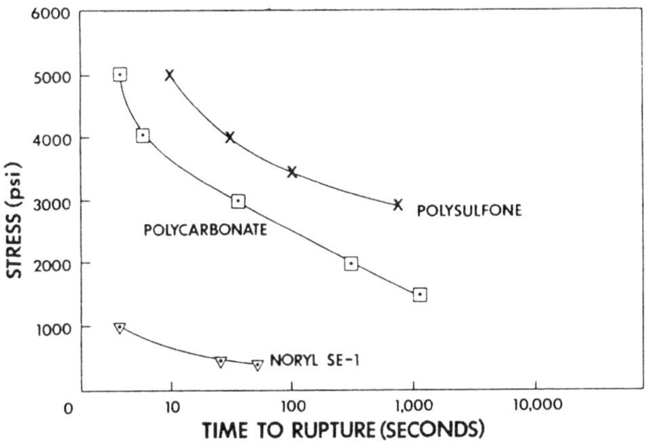

Fig. 6.3: ES rupture comparison of several engineering thermoplastics in a 70/30 isooctane/toluene mixture at 25°C [4]

6.1.3 Permeability

Plastics are to some extent permeable to gases, vapors and liquids. The diffusional characteristics of polymers can be described in terms of a quantity known as permeability.

The mass of the fluid permeating through the polymer at equilibrium conditions is given by [3]

$$m = \frac{P \cdot t \cdot A \cdot (p_1 - p_2)}{s} \tag{6.1.1}$$

where

m	=	mass of fluid permeating [g]
P	=	permeability $\left[\frac{g}{m \cdot s \cdot Pa}\right]$
t	=	time of diffusion [s]
A	=	Area of the film or membrane [m^2]
p_1 p_2	=	partial pressures on the side 1 and 2 of the film [Pa]
s	=	thickness of the film [m]

Besides its dependence on temperature the permeability is influenced by the difference in partial pressures of the fluid and thickness of the film. Other factors which affect permeability are the structure of the polymer film such as crystallinity and type of the fluid.

Permeability is a barrier property of plastics and is important in applications like packaging. Permeation data for some resins is summarized in Table 6.1 [3].

Table 6.1: Relative Permeability of Various Polymers Compared to PVDC [3]

Polymer	Gas permeability			
	N_2	O_2	CO_2	H_2O (25°C, 90% R.H.)
PVDC	1	1	1	1
PCTFE	3.19	10.64	2.48	0.21
PET	5.32	4.15	5.28	93
PA6	8.51	7.17	5.52	500
PVC-U	42.55	22.64	34.5	111
CA	298	147	235	5357
PE-HD	287	200	121	9.3
PE-LD	2021	1038	1214	57
PP	479	434	317	49
PS	309	208	303	857
Butyl Rubber	332	245	179	–
Polybutadiene	862	3604	4800	–
Natural Rubber	8600	4400	4517	–

Table 6.2: Relative Permeability to CO_2 of Different Polymers Compared to PET [3]

Polymer	Relative permeability
PET	1
PVC-U	6.53
PP	60
PE-HD	23
PE-LD	230

One of the reasons for using PET for making bottles for carbonated drinks is that PET is relatively impermeable to carbon dioxide, as can be seen from Table 6.2 [3].

6.1.4 Absorption and Desorption

The process by which a fluid is absorbed or desorbed by a plastics material is time-dependent, governed by its solubility and by the diffusion coefficient [3]. The period until equilibrium value is reached, can be very long. Its magnitude can be estimated by the half-life of the process given by [3]

$$t_{0.5} = \frac{0.04919\ s^2}{D} \tag{6.1.2}$$

where $t_{0.5}$ = half-life of the process
s = thickness of the polymer assumed to be penetrated by one side
D = diffusion coefficient

The value of $t_{0.5}$ for moisture in PMMA for $D = 0.3 \times 10^{-12}$ m^2/s and $s = 3$ mm is

$$t_{0.5} = \frac{0.04919 \cdot 3 \cdot 3 \cdot 10^{-6}}{0.3 \cdot 10^{-12} \cdot 3600 \cdot 24} = 17.1 \text{ days}$$

when the sheet is wetted from one side only [3]. However, the equilibrium absorption takes much longer, as the absorption rate decreases with saturation.

6.1.5 Weathering Resistance

Weathering is the deterioration of the polymer under atmospheric conditions which by the action of oxygen, temperature, humidity, and above all, ultra-violet radiation on the resin lead to loss of properties and finally failure of the part. Performance data of the resin under weathering is supplied by resin makers (see also Table 6.4) [8].

6.2 Chemical Resistance

Good chemical resistance of polymers is of importance in many applications. Table 6.3 shows the performance of some polymers according to a rating when they are in direct contact with the chemical agents listed in the Table 6.3 [6].

Table 6.3: Chemical Resistance of Resin to Different Liquid Agents [6]

Polymer	Salt solution	Concentrated acids	Organic acids	Concentrated bases	Aliphatic hydrocarbons	Halogenated hydrocarbons	Aromatic hydrocarbon	Monvalent alcohols	Oils and greases	Ketones	Esters	Esthers
PE-LD	1	2	1	1	1	5	3	3	2	3	3	3
PE-HD	1	2	1	1	1	4	3	1	3	1	1	1
PVC-U	1	1	2	1	1	5	5	1	1	5	5	1
PMMA	1	3	5	3	1	5	5	5	1	5	5	5
PS	1	3	5	3	1	5	5	5	1	5	5	5
PA	1	5	4	5	1	4	1	1	1	1	1	1
PC	1	2	4	5	1	5	5	2	1	5	5	5
POM	1	5	1	1	1	1	2	1	1	3	5	1
PTFE	1	1	1	1	1	1	1	1	1	1	1	1

Rating: 1: resistant; 2: sufficiently resistant; 3: conditionally resistant; 4: mostly not resistant; 5: completely nonresistant

Data on chemical resistance of resins can be obtained from resin manufacturers.

6.2.1 Chemical and Wear Resistance to Polymers

As polymers are converted into products they contact the walls of the various machines in which they are converted. Such walls include barrel walls, the plasticating screw surfaces, the interior die surfaces and the mold cavity surfaces. It is important that the correct materials be chosen for these surfaces or reduced performance and premature failure may arise. In the case of screws and barrels, the contacting surfaces must also exhibit wear resistance to each other.

One of the approaches is to select the barrel which provides the greatest resistance to corrosion and wear by particular plastic, including any additives, fillers or reinforcements. Then to select a material for the screw flights which performs well against that particular barrel material, and finally to select a coating for the remaining surfaces of the screw which exhibits the greatest corrosion and wear resistance to the plastic.

6.3 General Property Data

General property data for some plastics is presented in Tables 6.5 to 6.11. Some properties of the reinforced plastics are illustrated in Fig. 6.4 [5]. The comparability of different test methods is discussed in Table 6.12.

Table 6.4: Ultraviolet, Thermal and Biological Resistance of Some Plastics [8]

Polymer	Service temperature range °C	Ultraviolet resistance	Fungus resistance
Acetals	90/104	fair	excellent
Polyamides	-50/150	poor	excellent
PC	120	excellent	excellent
Polyethylene	75/100	poor	poor
Polypropylene	110	poor	excellent
PS	75	fair	excellent
Polysulfone	170	fair	excellent
PVC	55	poor	poor
Epoxy, aromatic	100/150	fair to good	poor
Epoxy, cycloaliphatic	100/150	excellent	poor
Polyurethanes	100/130	fair	poor
Silicones	260	excellent	excellent
Butyl rubber	-45/150	excellent	excellent

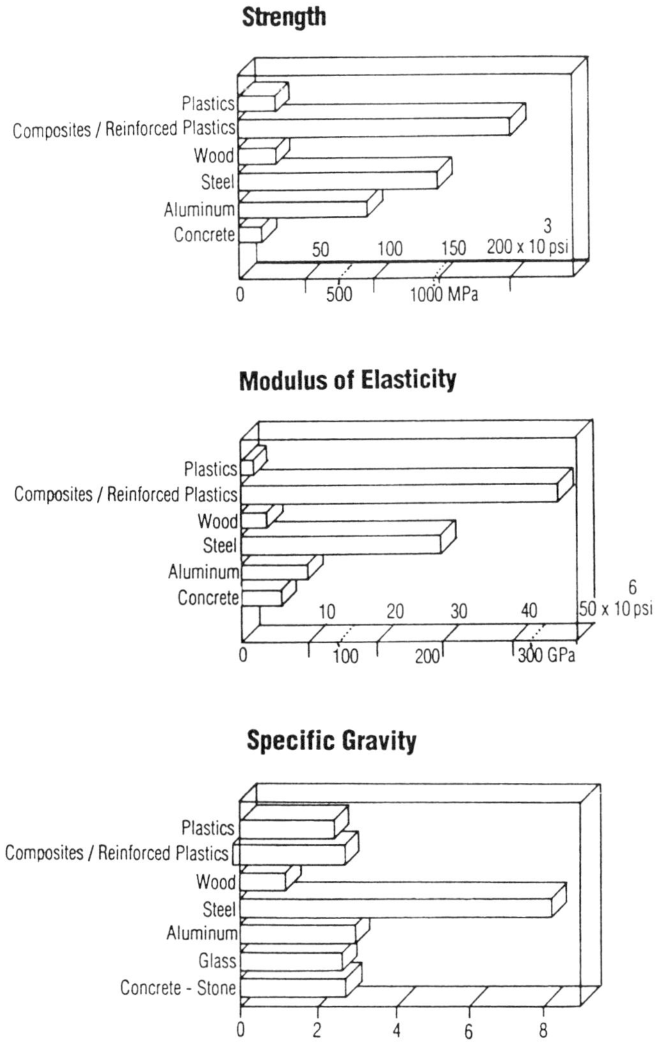

Fig. 6.4: Reinforced plastics and composites vs. other materials [5]

Table 6.5: Polymer Parameters and Their Influence Properties [8]

	Density		Molecular weight		Molecular weight distribution	
	Increases	Decreases	Increases	Decreases	Broadens	Narrows
Environmental stress	↓	↑	↑	↓	↑	↓
Impact strength	↓	↑	↑	↓	↓	↑
Stiffness	↑	↓	–	–	–	–
Hardness	↑	↓	–	–	–	–
Tensile strength	↑	↓	–	–	–	–
Permeation	↓	↑	–	–	–	–
Warpage	↑	↓	–	–	↑	↓
Abrasion resistance	–	–	↑	↓		
Flow processibility	–	–	↓	↑	↑	↓
Melt strength	–	–	↑	↓	↑	↓
Melt viscosity	–	–	↑	↓	↓	↑
Copolymer content	↓	↑	–	–	–	–

Table 6.6: Properties of Some Thermoplastic Structural Foams (Physical Properties Guidelines at 0.25 Wall with 20% Density Reduction) [9]

Property	PE-HD	ABS	modified PPO	PC
Specific gravity	0.6	0.86	0.85	0.9
Deflection temperature [°C]				
under load at 0.462 N/mm^2	54.2	86	96	138
at 1.85 N/mm^2	34.2	78	82	127
Coefficient of thermal expansion [$K^{-1} \cdot 10^{-5}$]	22	8.9	6.9	3.64
Tensile strength [N/mm^2]	9.2	27.3	23.8	42.7
Flexural modulus [N/mm^2]	840	1960	1827	2500
Compressive strength 10% deformation [N/mm^2]	12.9	31	36.4	36.4

Table 6.7: General Properties of Liquid-Crystal Polymers [8]

Property	Value
Flexural strength, N/mm^2	147/308
Unnotched izod, J/m	150/300
Notched izod, J/m	400/500
Rockwell hardness	M62/99
Dielectric constant, 10^3 Hz	2.9/4.5
Dissipation factor, 10^3 Hz	$4–6\times10^{-3}$
Volume resistivity, W · cm	$10^{12}/10^{-13}$
Dielectric strength, V/mil	780/1000
Arc resistance, s	63/185
Water absorption, equilibrium at 23 °C, %	0.02/0.04
Specific gravity	1.4/1.9
Glass transition temperature, °C	–
Melt temperature, °C	275/330
Heat distortion temperature, °C	
at 0.455 N/m^2	250/280
at 1.848 N/m^2	180/240
Continuous-use temperature, °C	200/240
Coefficient of thermal expansion, ppm/°C	$0–25\times10^{-6}$ flow direction $25–50\times10^{-6}$ transverse
Flammability, UL-94	V/O
Oxygen index	35/50
Tensile modulus, N/mm^2	9.8/40.6
Tensile strength, yield N/mm^2	140/245
Tensile elongation, %	1.2/6.9
Flexural modulus, N/mm^2	9.8/35

Table 6.8: Some Physical Properties of Polyurethane Structural Foams [9]

Property vs specific gravity			
Specific gravity	0.4	0.5	0.6
Flexural strength [N/mm^2]	24.5	32.2	36.4
Flexural modulus [N/mm^2]	602	742	847
Tensile strength [N/mm^2]	9.8	12.6	16.8
Heat deflection temperature [°C] at 0.462 n/mm^2	81	91	96
Charpy impact, unnotched [kT/m^2]	13.25	18.1	21
Skin hardness, D scale	70	75	80
Property vs thickness			
Thickness	1/4in	3/8in	1/2in
Specific gravity	0.6	0.6	0.6
Flexural strength (N/mm^2)	37.8	37.8	36.4
Flexural modulus (N/mm^2)	1092	1001	847
Tensile strength (N/mm^2)	18.2	17.5	16.8
Charpy impact, unnotched (ft · lb/in^2)	12.0	11.0	10.0

Table 6.9: Properties of Some Thermosetting Resins [8]

	Epoxy		general	Phenolics	
	glass filled	mineral filled	purpose	glass filled	mineral filled
Specific gravity	1.8	2.1	1.45	1.95	1.83
Water absorption [% 24h]	0.2	0.04	0.7	0.5	0.5
Heat deflection temperature at 1.85 N/m^2 [°C]	204	121	172	204	260
Tensile strength [N/mm^2]	210	105	70	49	77
Impact strength [T/m)] [Izod]	530.4	21.4	16	187	801
Coefficient of thermal expansion [$10^{-5} \cdot K^{-1}$]	3.1	4	4.55	–	1.76
Thermal conductivity [W/m · k]	0.864	–	0.043	0.05	0.03

Table 6.10: Properties of Some Elastomers [8]

Property	EPM/EPDM	Chloroprene	Fluoro-elastomers	Natural rubber
Tensile strength (N/mm^2)	3.5/24.5	3.5/24.5	10.5	16.8/32.2
Tensile modulus (N/mm^2)	700/21000	700/21000	1400/14000	3360/5950
Elongation	100/700	100/800	150/450	300/750
Resilience	good	excellent	fair	outstanding
Brittleness temperature (°C)	-55	-45	-15/-50	-60
Thermal expansion coefficient ($10^{-5} \cdot K^{-1}$)	58	62	55	67
Durometer hardness	30A/90A	30A/95A	55A/95A	30A/100A
Compression set	20/60	20/60	15/30	10/30

Table 6.11: Comparative Physical Properties of Metals and Reinforced Plastics at Room Temperature [10]

Property	Carbon steel 1020	Stainless steel 316	Hastelloy C	Aluminium	Glassmat laminate	Composite structure glassmat woven roving	Glass reinforced epoxy filament wound
Density [kg/m^3]	7860	7916	8968	2712	1384	1522	1800
Coefficient of thermal expansion [$k^{-1} \cdot 10^{-6}$]	11.83	16.7	11.5	24	31	24	16.4/22
Modulus of elasticity [N/mm^2]	210000	196000	196000	70000	4900/7000	7000/10500	21000
Tensile strength [N/mm^2]	462	595	560	84	63/84	84/140	420/700
Yield strength [N/mm^2]	231	245	354	28	63/84	84/140	420/700
Thermal conductivity [W/m · k]	48.5	16.3	11.3	234	2.6	2.6	2.6/3.5
Ratio of tensile strength to density	0.0588	0.0752	0.0624	0.031	0.0607	0.092	0.39

Table 6.12: Comparability of DIN/ISO and ASTM Test Methods for Plastics [2]

Test	Standard	Symbol	Units	Comparability
Density	DIN1306 DIN53479 ASTM D 792 ASTM D 1505	d d_R d^{23C}	g/cm3 at 20°C g/ml at 23°C	comparable at same temperature
Melt index	DIN 53735 (~ISO/R 292) ASTM D 1338	MFI –	g/10 min g/10 min	comparable under same conditions

(cont. next page)

Test	Standard	Symbol	Units	Comparability
Mechanical properties				
Elastic modulus	DIN 53457	E	kN/mm2	comparable only under same
	ASTM D 638	E	psi	conditions: shape and state
	(Tensile)		(=lb/sq.in.)	of specimen, measured
	ASTM D 695	–		lenght, test rate
	(Compression)			
	ASTM D 790	E_B	psi	
	(Fluxural)			
	ASTM D 882	–	kgf/cm^2 *)	
	(Film)			
Shear modulus	DIN 53445	G	kN/mm^2	
	(ISO/DR 533)			comparable
Torsion modulus	ASTM D 2236	G	dyn/cm^2	
			psi	
Torsional stiffness	DIN 53447	T	kN/mm^2	
	(ISO/DR 458)			comparable
Stiffness properties	ASTM D 1043	G (!)	dN/cm^2	
Tensile properties				
Tensile strength	DIN 53455	σ_B	N/mm^2	
(at maximum load)	(ISO/DR 468)			
	DIN 53371	σ_B	N/mm^2	
	ASTM D 638	–	psi	comparable only under same
	ASTM D 882	–	dN/cm^2	conditions:
Elongation at maximum	DIN 53455	ε_B	%	shape and state of specimen,
force	DIN 53371	δ_{Pmax}	%	measured length, test rate
Percentage elongation	ASTM D 638	%El	%	
	ASTM D 882	–	%	
Ultimate tensile strength	DIN 53455	σ_R	N/mm^2	
	DIN 53371	σ_R	N/mm^2	
Tensile strength	ASTM D 638	σ_U	psi	
(at break)	ASTM D 882	–	dN/cm^2	
Percentage elongation	DIN 53455	δ_R	%	
(at break)	DIN 53371	δ_R	%	
	ASTM D 638	–	%	
	ASTM D 882	–	%	
Yield stress	DIN 53455	σ_S	N/mm^2	comparable only under same
(Yield point)	DIN 53371	σ_S	N/mm^2	conditions:
Yield point	ASTM D 638	–	psi	shape and state of specimen,
Yield strength	ASTM D 882	–	dN/cm^2	measured length, test rate
Percentage	DIN 53455	E_S	%	
Elongation (at yield)	ASTM D 882	–	%	
Flexural properties	DIN 53452	σ_{bB}	N/mm^2	
Flexural strength	ASTM D 790	–	psi	
Limiting flexural stress	DIN 53452	σ_{bG}	N/mm^2	limited comparability
Flexural stress	ASTM D 790	–	psi	
(at 5% strain or at				
conventional deflection)				
Flexural impact	DIN 53453	a_n	kJ/m^2	not comparable
Impact resistance	(ISO/R 179)			
Impact strength	ASTM D 256	–	J/m	
	(ISO/R 180)			

*) kgf = kiloforce = dN

Test	Standard	Symbol	Units	Comparability
Impact strength (notched)	DIN 53453 (ISO/R 179) ASTM D 256 (ISO/R 180)	a_K –	kJ/m² J/m	not comparable
Shore hardness Durometer (Shore hardness)	DIN 53505 ASTM D 1706 ASTM D 2240	– – H	Shore scale A, C, D Scale A, D gramforce	comparable
Ball indention hardness α-Rockwell hardness	DIN 53456 ASTM D 785	H –	N/mm² Rockwell scale	not comparable
Thermal properties				
Vicat softening point	DIN 53460 (ISO/R 306) ASTM D 1525	VSP –	°C °C	comparable if measured under same load in liquid
Martens heat distortion temperature Heat distortion temperature Deflection temperature (HDT)	DIN 53458 DIN 53461 (ISO/R 75) ASTM D 648	– F_{ISO} –	°C °C °C, °F	comparable if load, specimen shape and slate are the same
Water absorbtion	DIN 53472 DIN 53475 (ISO/R 62) ASTM D 570	– –	22°C, 4d, mg 23°C, 24h, mg 23°C, 24h, % 2h	comparable (converted), if pre- and post-treatment, specimen and times are the same
Electrical properties				
Volume resistivity	DIN 53482 (VDE 0303 Part 3) ASTM D 257	ρ_D ρ	Ω·cm Ω·cm	comparable
Surface resistivity	DIN 53482 (VDE 0303 Part 3) ASTM D 257	R_0 σ	Ω Ω	comparable
Dielectrical constant	DIN 53483 (VDE 0303 Part 4) ASTM D 150	ε_r ε	– –	comparable
Dissipation factor (tan δ)	DIN 53483 (VDE 0303 Part 4) ASTM D 150	– D–	–	comparable
Dielectrical loss factor (ε= tan δ)	– ASTM D 150	– –	– –	
Dielectrical strength	DIN 53481 (VDE 0303 Part 2) ASTM D 149	E_d –	kV/mm V/mil (1 mil=25µm)	not comparable
Arc resistance	DIN 53484 (VDE 0303 Part 5) ASTM D 495	– –	sec sec	not comparable
Tracking resistance	DIN 53480 (VDE 0303 Part 1) ASTM D 2132	– –	– –	not comparable

Literature

1. *Birley, A.W.: Haworth, B.: Bachelor, T.:* Physics of Plastics, Hanser, Munich 1991
2. *Domininghaus, H.:* Plastics for Engineers, Hanser, Munich 1993
3. *Ogorkiewicz, R.M.:* Thermoplastics Properties and Design, John Wiley, New York 1973
4. *Harris, J.E.:* Polysulfone in Engineering Thermoplastics: Properties and Applications. Ed.: James M. Margolis, Marcel Dekker, Basel 1985
5. *Rosato, D.V., Rosato, D.V.:* Plastics Processing Data Handbook, Van Nostrand Reinhold, New York 1990
6. *Schmiedel, H.:* Handbuch der Kunststoffprüfung (Ed.), Hanser, Munich 1992
7. *Maresca, L.M., Robeson, L.M.:* Polyarylates in Engineering Thermoplastics, Properties and Applications. Ed. James M. Margolis, Marcel Dekker, Basel, 1985
8. *Alvino, W.M.:* Plastics for Electronics - Materials, Properties and Design Applications, Mcgraw Hill, New York 1994
9. *Wendle, B.A.:* Structural Foam, A purchasing and Design Guide, Marcel Dekker, New York 1985
10. *Mallinson, J.H.:* Corrosion-resistant Plastic Composites in Chemical Plant Design, Marcel Dekker, New York 1969

7 Extrusion

Extrusion is one of the most widely used unit operations of polymer processing. Basically it consists of transporting the solid polymer in an extruder by means of a rotating screw, melting the solid, homogenizing the melt, and forcing the melt through a die (Fig. 7.1).

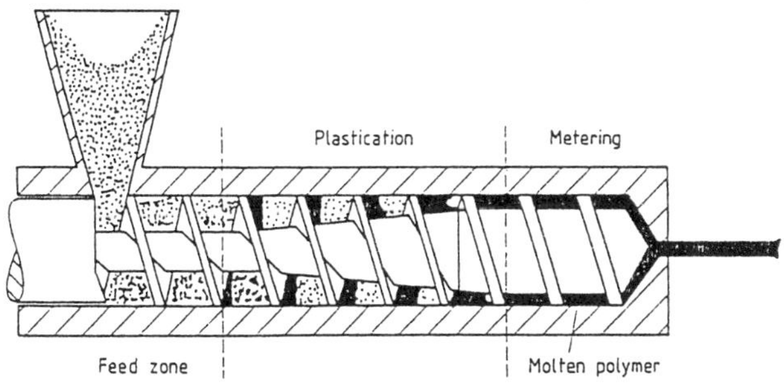

Fig. 7.1: Plasticating extrusion [2]

The extruder screw of a conventional plasticating extruder has three geometrically different zones (Fig. 7.1a) whose functions can be described as follows:

feed zone
transport and preheating of the solid material
transition zone
compression and plastication of the polymer
metering zone
melt conveying, melt mixing and pumping of the melt to the die

However the functions of a zone are not limited to that particular zone alone. The processes mentioned can continue to occur in the adjoining zone as well.

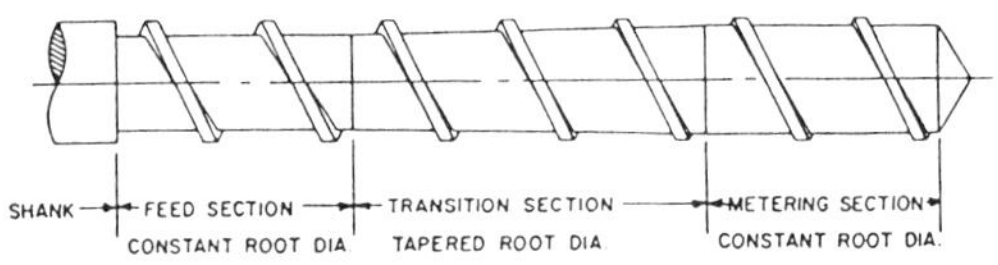

Fig. 7.1a: Three zone screw

7.1 Extrusion Screws

To perform these processing operations optimally screws of varied geometry as shown in Fig. 7.2 are used. Higher melting capacities compared to the three zone screws (Fig. 7.1) are achieved with screws having shearing and mixing devices (Fig. 7.2a). Barrier type screws enable the separation of solids polymer from the melt, and thus lead to lower and constant melt temperatures over a wide range of screw speeds (Figs. 7.2c/d). Devolatilizing screws are applied to extract volatile components from the melt (Fig. 7.2b) [27].

Twin screw extruders are mainly used in the compounding of polymers, such as, for example, in the pelletizing and compounding of PVC or in the profile extrusion of PVC. The positive conveying characteristics of the twin screw extruders are achieved by forcing the material to move in compartments formed by the two screws and the barrel. The degree of intermeshing between the flight of one screw and the channel of the other screw, sense of rotation and speed distinguish the different kinds of screws (Fig. 7.3 [9]) from one another.

7.2 Processing Parameters

The factors affecting extrusion can be classified into resin-dependent and machine related parameters.

7.2.1 Resin-Dependent Parameters

Resin-dependent parameters are not the physical constants which one obtains by measuring the physical properties of the polymeric material. They are, however, similar to these constants with the difference that they relate more to the process and processing conditions. If, for example, the processing temperature of a polymer is too high, degradation of the material leading to adverse material properties of the product might occur. Too low a processing temperature on the other hand may create a melt containing unmelted solids. The appropriate processing temperatures and pressures suited to a particular resin and process are usually quoted by the resin manufacturers. These values can be maintained during the process by optimizing the processing conditions and, if necessary, the machinery such as extrusion screw or die [1].

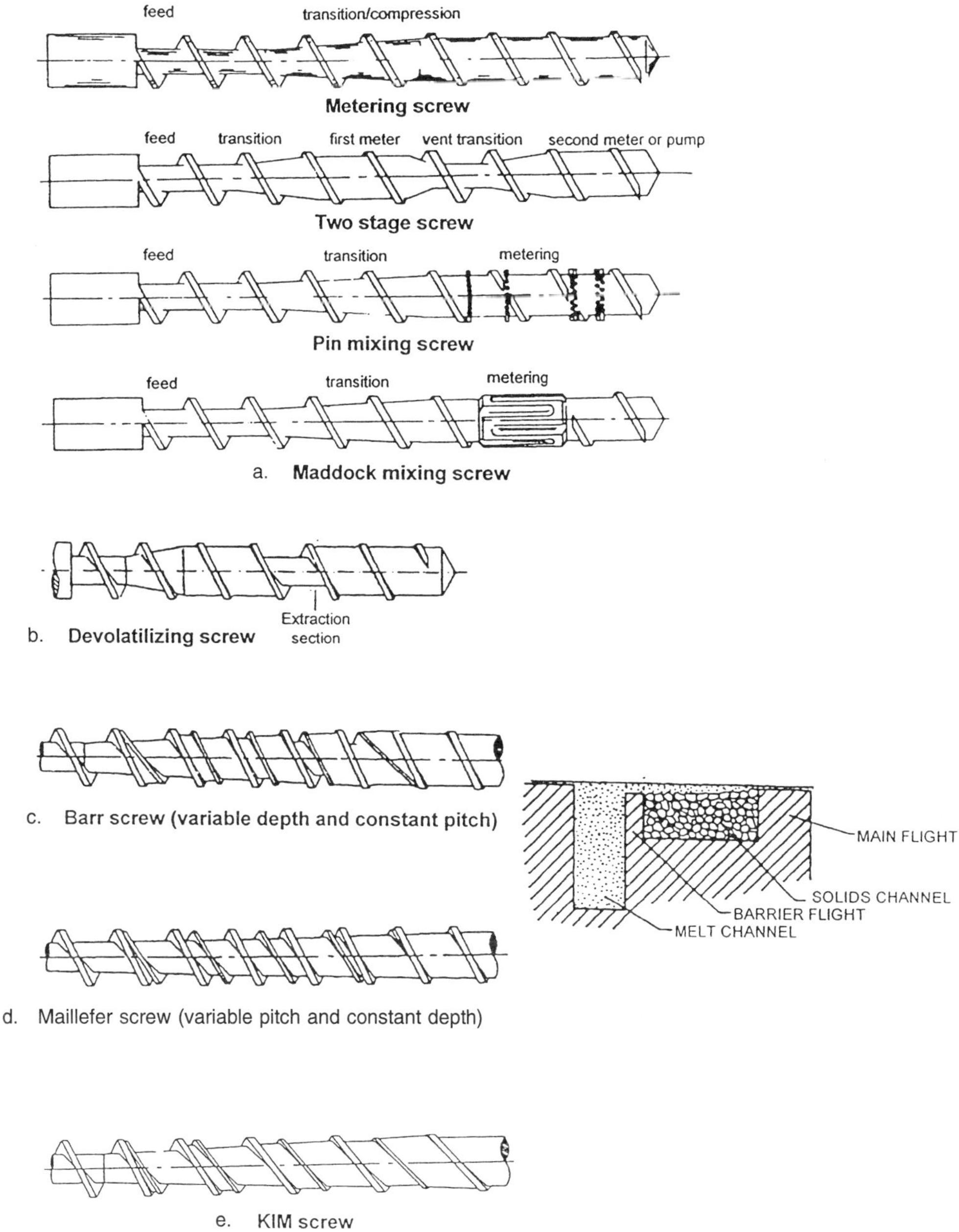

Fig. 7.2: Geometries of some extrusion screws [20], [27]

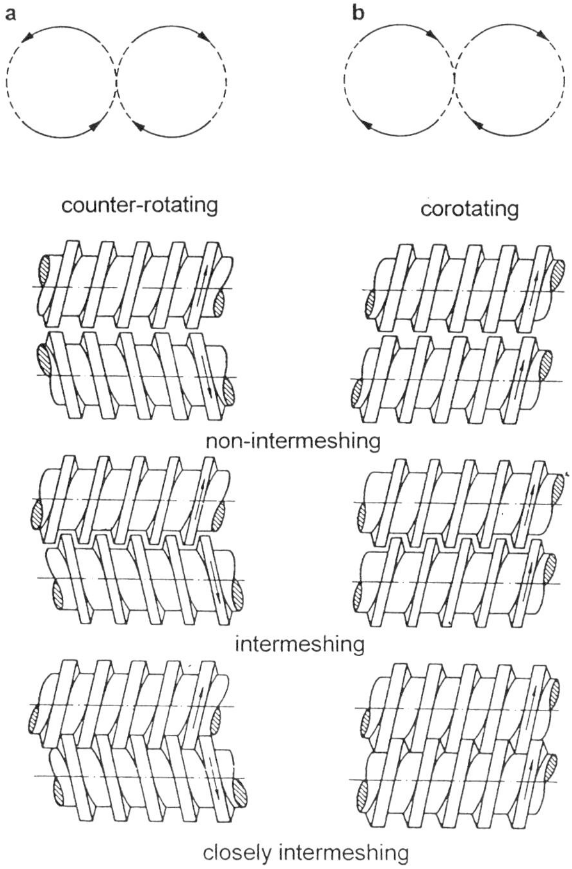

Fig. 7.3: Geometries of some twin screws [9]

Melt temperature and pressure are important resin-dependent parameters which affect the quality of the product. Guide values of these parameters are given below for a few resins and extrusion processes. They refer to the temperature and pressure of the melt at the die entrance as shown in Fig. 7.4. These values are also known in the practice as stock temperature and stock pressure, and depend not only on the type of the resin but also on the grade of the resin used, which is characterized, for example, by the MFR value.

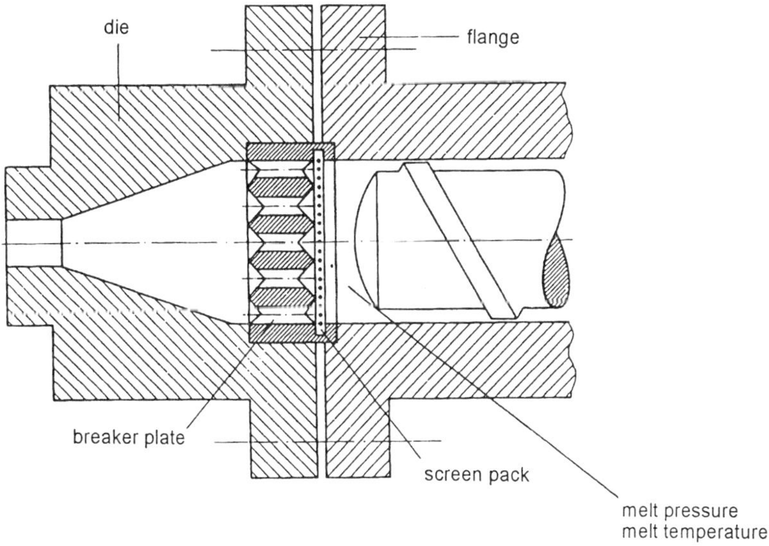

Fig. 7.4: Position of measurement of melt pressure and melt temperature [9]

Blown Film (Fig. 7.5)

Characteristic values of melt temperature and pressure are given in Table 7.1.

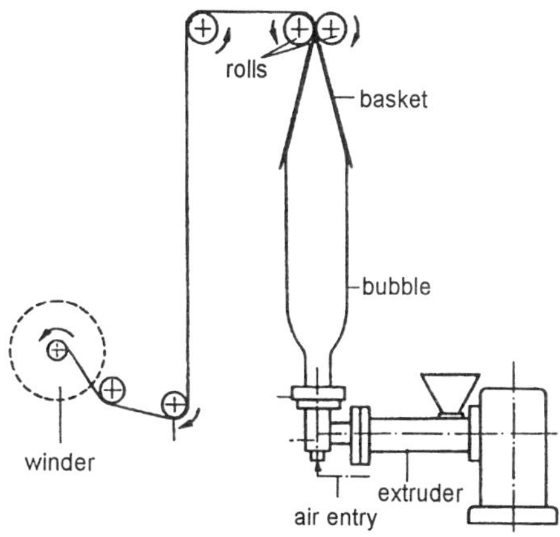

Fig.7.5: Blown film [9]

Table 7.1: Typical Values of Melt Temperature and Pressure for Blown Film

Material	Melt temperature °C	Melt pressure bar
PE-LD	140/210	100/300
PE-HD	180/230	150/300
PE-LLD	190/230	250/350
PP	220/240	200/300
PA 6	270/280	160/200
PC	260/290	150/250
PET	260/280	150/250
EVA	150/200	100/200
EVOH	200/220	100/200

Pipe Extrusion (Fig. 7.6)

Table 7.2 shows some guide values of melt temperature and pressure for pipe extrusion.

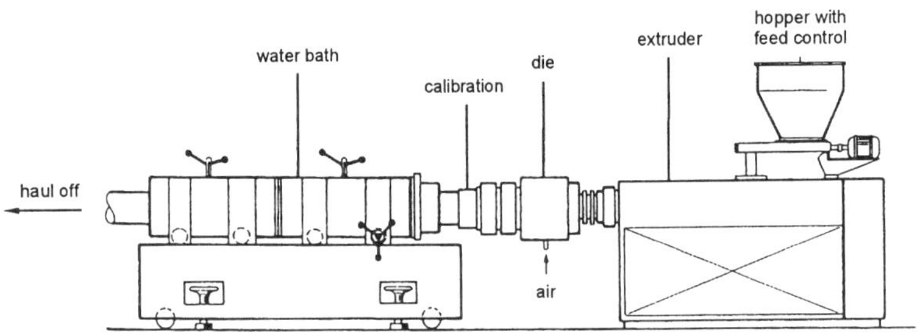

Fig. 7.6: Pipe extrusion [9]

Table 7.2: Typical Values of Melt Temperature and Melt Pressure for Pipe Extrusion

Material	Melt temperature °C	Melt pressure bar
PVC-U	180/185	100/200
PVC-P	160/165	100/200
PEHD	180/200	150/250
PP	230/240	150/200

Flat Film Extrusion (Fig. 7.7)

Typical values of melt temperature and pressure for flat film extrusion are given in Table 7.3.

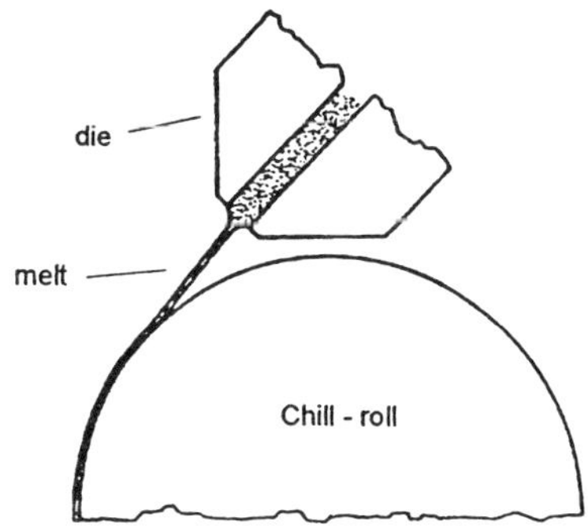

Fig. 7.7: Flat film extrusion [9]

Table 7.3: Typical Values of Melt Temperature and Pressure for Flat Film Extrusion

Material	Melt temperature °C	Melt pressure bar
PP	230/260	150/200
PA 6	260/280	100/200
PELD	220/250	100/250
PEHD	220/230	150/250

Sheet Extrusion

A flat die is used in both flat film and sheet extrusion processes. Depending on the thickness of the product the term film or sheet is applied. Typical values for sheet extrusion are presented in Table 7.4.

Wire Coating

Melt temperatures for wire coating are approximately the same as those used for pipe extrusion. However, depending on the type of extrusion the pressure can range from 100 bar to 400 bar.

Melt temperatures and pressures for other resins and processes can be obtained from resin makers. Data on processing conditions suited to attain resin compatible melt temperatures is also supplied by resin manufacturers. This data includes values on extruder barrel temperature, die temperature, screw speed, and depending on the process haul-off rates and draw-ratios.

Table 7.4: Typical Values of Melt Temperature and Pressure for Sheet Extrusion

Material	Melt temperature °C	Melt pressure bar
PS	190/220	150/250
PMMA	200/230	200/250
SB	210/230	150/250
ABS	220/240	150/250
PET	280/285	150/200
PVC-U	180/185	100/200
PVC-P	160/165	75/150

7.2.2 Machine Related Parameters

Although machine related parameters include effects due to resin properties, they are more influenced by the geometry of the machine elements involved such as extrusion screw and die than by resin properties. Following examples taken from single screw extrusion illustrate the influence of the geometry of machinery on the target quantities of the process.

7.2.2.1 Extruder Output

Depending on the type of extruder the output is determined either by the geometry of the solids feeding zone alone as in the case of a grooved extruder [11] or by the solids and melt zones to be found in a smooth barrel extruder.

Feed Zone

The solids transport is largely influenced by the frictional forces between the solid polymer and barrel and screw surfaces. A detailed analysis of the solids conveying mechanism was performed by Darnell and Mol [16]. In the following example an empirical equation which gave good results in the practice is presented [1], [2].

Example:

The geometry of the feed zone of a screw (Fig. 7.8), is given by the following data:

barrel diameter	D_b	=	30 mm
screw lead	s	=	30 mm
number of flights	ν	=	1
flight width	w_{FLT}	=	3 mm
channel width	W	=	28.6 mm

depth of the feed zone	H	=	5 mm
conveying efficiency	η_F	=	0.436
screw speed	N	=	250 rpm
bulk density of the polymer	ρ_o	=	800 kg/m^3

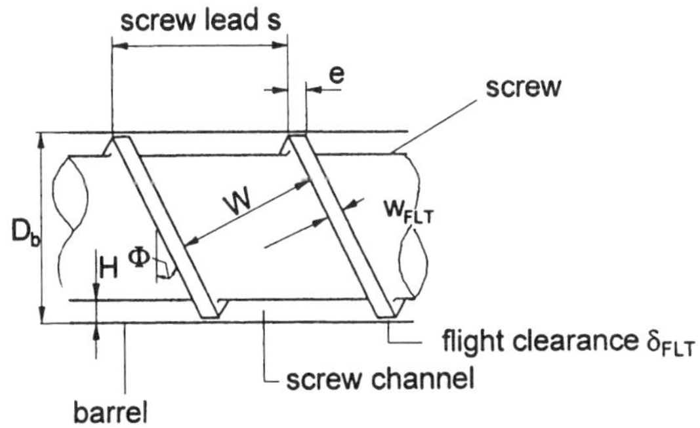

Fig. 7.8: Screw zone of a single screw extruder

The solids conveying rate in the feed zone of the extruder can be calculated according to [2]

$$G = 60 \cdot \rho_o \cdot N \cdot \eta_F \cdot \pi^2 \cdot H \cdot D_b\ (D_b - H)\ \frac{W}{W + w_{FLT}} \cdot \sin\phi \cdot \cos\phi \qquad (7.2.1)$$

with the helix angle Φ

$$\phi = \tan^{-1}[s/(\pi \cdot D_b)] \qquad (7.2.2)$$

The conveying efficiency η_F in Eq. (7.2.1) as defined here is the ratio between the actual extrusion rate and the theoretical maximum extrusion rate attainable under the assumption of no friction between the solid polymer and the screw. It depends on the type of polymer, bulk density, barrel temperature, friction between the polymer, barrel and the screw. Experimental values of η_F for some polymers are given in Table 7.4A.

Solution:

Inserting the values above with the dimensions in meters into Eq. (7.2.1) and Eq. (7.2.2) we get

$$G = 60 \cdot 800 \cdot 250 \cdot 0.44 \cdot \pi^2 \cdot 0.005 \cdot 0.03 \cdot 0.025 \cdot \frac{0.0256}{0.0286} \cdot 0.3034 \cdot 0.953$$

Hence G ≈ 50 kg/h

Table 7.4A: Conveying Efficiency η_F for Some Polymers

Polymer	Smooth barrel	Grooved barrel
PE-LD	0.44	0.8
PE-HD	0.35	0.75
PP	0.25	0.6
PVC-P	0.45	0.8
PA	0.2	0.5
PET	0.17	0.52
PC	0.18	0.51
PS	0.22	0.65

Metering Zone (Melt Zone)

Starting from the parallel plate model and correcting it by means of appropriate correction factors [3] the melt conveying capacity of the metering zone can be calculated as follows [3]:

Although the following equations are valid for an isothermal quasi-Newtonian fluid, there were found to be useful for many practical applications [1].

These equations can be summarized as follows:

Volume flow rate of pressure flow $\dot{Q}_p$ (Fig. 7.8A):

$$\dot{Q}_p = \frac{-\pi D_b \cdot H^3 \left(1 - \frac{\nu \cdot e}{s}\right) \cdot \sin^2\phi \cdot \Delta p \cdot 10^{-4}}{12\,\eta_a \cdot L} \qquad (7.2.3)$$

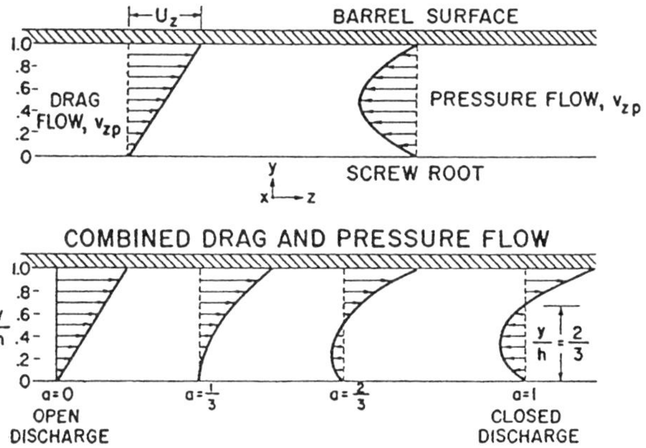

Fig. 7.8A: Drag and pressure flow in screw channel

Mass Flow Rate $\dot{m}_p$

$$\dot{m}_p = 3600 \cdot 1000 \cdot \dot{Q}_p \cdot \rho_m \qquad (7.2.4)$$

Drag Flow $\dot{Q}_d$ *(Fig. 7.8A)*

$$\dot{Q}_d = \frac{\pi^2 \cdot D_b^2 \cdot N \cdot H \cdot \left(1 - \frac{\nu \cdot e}{s}\right) \cdot \sin\phi \cdot \cos\phi \cdot 10^{-9}}{120} \qquad (7.2.5)$$

Mass Flow Rate $\dot{m}_d$

$$\dot{m}_d = 3600 \cdot 1000 \cdot \dot{Q}_d \cdot \rho_m \qquad (7.2.6)$$

Leakage Flow $\dot{Q}_L$

To avoid metal to metal friction, extrusion screws have a small clearance between the top of the flight and the barrel surface. This clearance reduces the pumping rate of the melt because it enables the polymer to leak across the flights. The net flow rate $\dot{Q}$ is therefore

$$\dot{Q} = \dot{Q}_d + \dot{Q}_p - \dot{Q}_L \qquad (7.2.7)$$

The melt conveying rate of the metering zone can be finally calculated from [1]

$$\dot{m} = 3 \cdot 10^{-5} \cdot \pi^2 \cdot D_b^2 \cdot N \cdot H \cdot \left(1 - \frac{\nu \cdot e}{s}\right) \cdot \rho_m \cdot \sin\phi \cdot \cos\phi \cdot (1 - a_d - J) \qquad (7.2.8)$$

where $a_d = -\dot{m}_p / \dot{m}_d$ and $J = \delta_{FLT} / H$

Average shear rate in the metering channel

$$\dot{\gamma}_a = \pi \cdot D_b \cdot N / 60 \cdot H \qquad (7.2.9)$$

Symbols and units used in the equations above

D_b	:	Barrel diameter [mm]
H	:	Channel depth [mm]
e	:	Flight width [mm]
s	:	Screw lead [mm]
δ_{FLT}	:	Flight clearance [mm]
L	:	Length of metering zone [mm]
$\dot{Q}_p$, $\dot{Q}_d$	:	Volume flow rate of pressure and drag flow, respectively [m^3/s]
$\dot{m}_p$, $\dot{m}_d$	:	Mass flow rate of pressure and drag flow respectively [kg/h]
$\dot{m}$	:	Extruder output [kg/h]

Δp	:	Pressure difference across the metering zone [bar]
ν	:	Number of flights
η_a	:	Melt viscosity [Pa · s]
$\dot{\gamma}_a$	:	average shear rate [s^{-1}]
ρ_m	:	density of the melt [g/cm^3]
a_d	:	Ratio of pressure flow to drag flow
N	:	Screw speed [rpm]

Example:
For the following conditions the extruder output is to be determined:
melt viscosity $\eta_a = 1406.34$ Pa·s; $N = 80$ rpm $\Delta p = 300$ bar; $\rho_m = 0.7$ g/cm^3
Geometry of the metering zone (Fig. 7.8)

$D_b = 60$ mm; $H = 3$ mm; $e = 6$ mm; $s = 60$ mm; $\delta_{FLT} = 0.1$ mm;
$L = 600$ mm; $\nu = 1$

Substituting these values into the equations above one obtains

$$\dot{m}_p = 3.148 \text{ kg/h} \qquad \text{Eq. (7.2.4)}$$
$$\dot{m}_d = 46.59 \text{ kg/h} \qquad \text{Eq. (7.2.6)}$$
$$\dot{m} = 41.88 \text{ kg/h} \qquad \text{Eq. (7.2.8)}$$

Leakage flow $\dot{m}_l = \dot{m}_d + \dot{m}_p - \dot{m} = 1.562$ kg/h

With the help of Eq. (7.2.8) the effect of different parameters on the extruder output is presented in Figs. 7.9A-H by changing one variable at a time and keeping all other variables constant. The dimensions of the screw used in these calculations are taken from the example above.

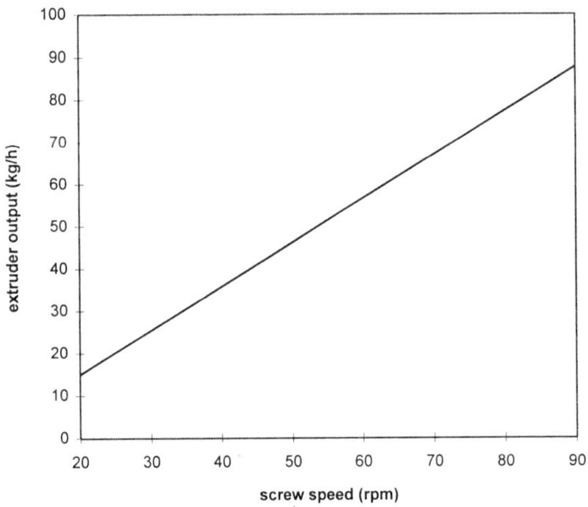

Fig. 7.9A: Effect of screw speed on extruder output

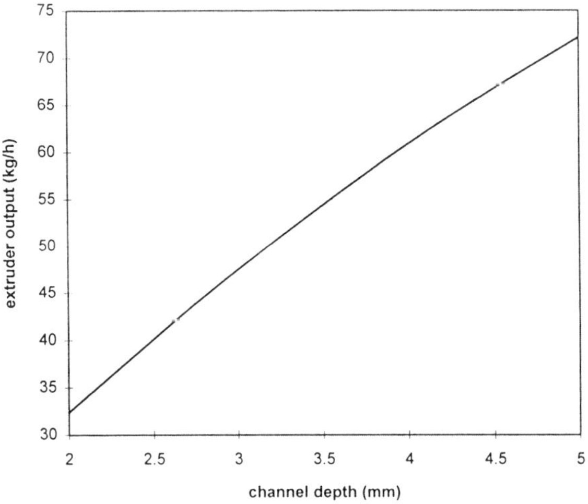

Fig. 7.9B: Effect of channel depth on extruder output

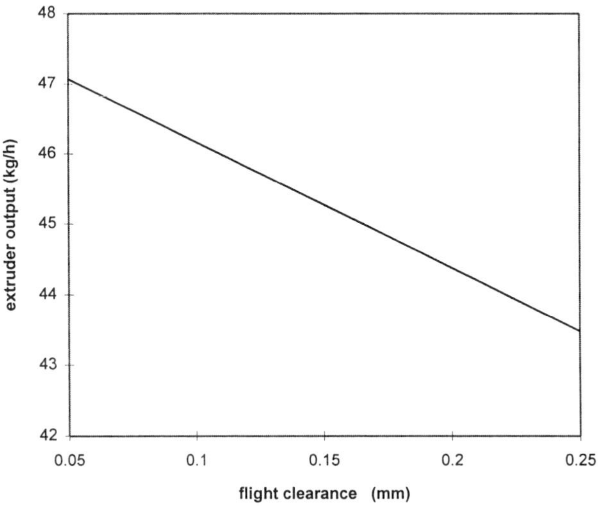

Fig. 7.9C: Effect of flight clearance on extruder output

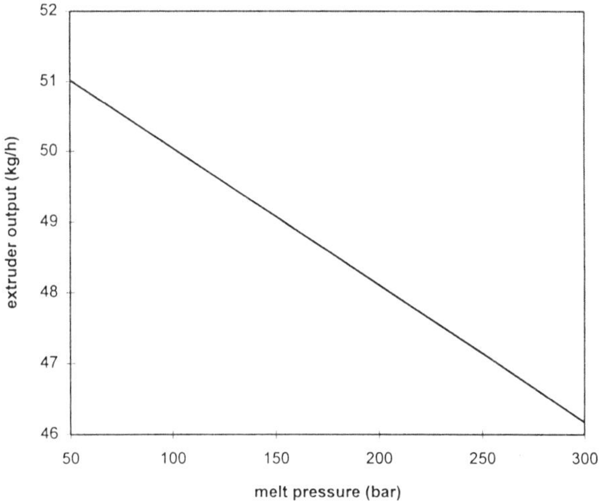

Fig. 7.9D: Effect of melt pressure on extruder output

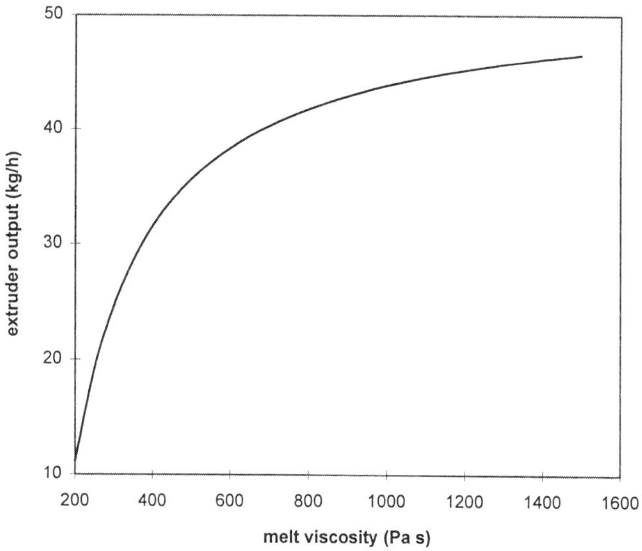

Fig. 7.9E: Effect of melt viscosity on extruder output

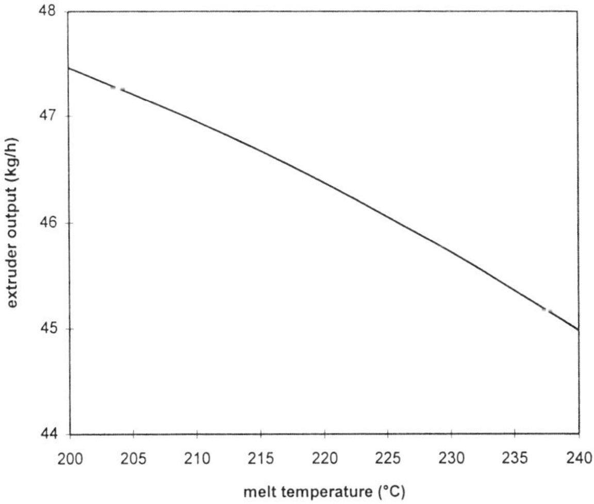

Fig. 7.9F: Effect of melt temperature on extruder output

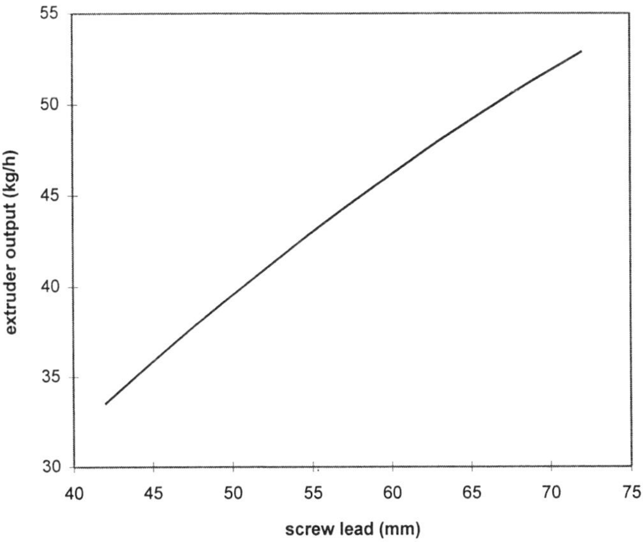

Fig. 7.9G: Effect of screw lead on extruder output

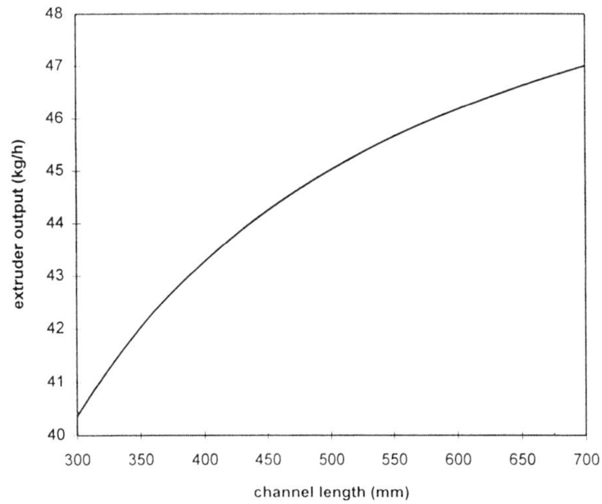

Fig. 7.9H: Effect of channel length on extruder output

Correction Factors

To correct the infinite parallel plate model for the flight edge effects following factors can be used along with the equations above:

With sufficient accuracy the shape factor F_d for the drag flow can be obtained from [27]

$$F_d = 1 - 0.571 \frac{H}{W} \tag{7.2.10}$$

and the factor F_p for the pressure flow

$$F_p = 1 - 0.625 \frac{H}{W} \tag{7.2.11}$$

The expressions for the corrected drag flow and pressure flow would then be

$$\dot{Q}_{dk} = F_d \cdot Q_d \tag{7.2.12}$$

and

$$\dot{Q}_{pk} = F_p \cdot \dot{Q}_p \tag{7.2.13}$$

The correction factor for the screw power which is treated in the next section can be determined from [1]

$$F_x = e^x - x^3 + 2.2\,x^2 - 1.05x \tag{7.2.14}$$

with $x = H/W$.

Eq. (7.2.14) is valid in the range $0 < H/W < 2$. For the range of commonly occurring H/W-ratios in extruder screws the flight edge effect accounts for only less than 5% and can therefore neglected [27]. The influence of screw curvature is also small so that F_x can be taken as 1. Although the above mentioned factors are valid only for Newtonian fluids; their use for polymer melt flow is accurate enough for practical purposes.

7.2.2.2 Melting Parameter

The melting rate is described by TADMOR [2] through the parameter Φ_p which is expressed as [2], [1] (see also Fig. 7.10).

$$\phi_p = \left\{ \frac{V_{bx} \cdot \delta_m \cdot \left[\lambda_m (T_b - T_m) + 0.5\, \eta_f \cdot V_j^2 \cdot 10^{-4}\right]}{200 \left[c_{p_s} (T_m - T_s) + i_m\right]} \right\}^{0.5} \tag{7.2.15}$$

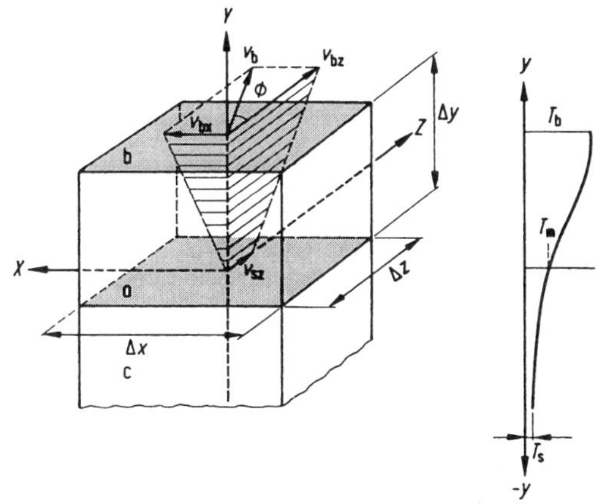

Fig. 7.10: Velocity and temperature profiles in the melt and solid bed after TADMOR [2]

The numerator represents the heat supplied to the polymer by conduction through the barrel and dissipation, where as the denominator shows the enthalpy required to melt the solid polymer. The melting rate increases with increasing Φ_p. The dimensionless melting parameter ψ is defined as [2]

$$\psi = \frac{\phi_p \cdot H_1 \cdot W_{0.5}}{10^{1.5} \cdot G} \tag{7.2.16}$$

This parameter is the ratio between the amount of melted polymer per unit down channel distance to the extruder output per unit channel feed depth.

The symbols and units used in the Eqns (7.2.15) and (7.2.16) are as follows:

Barrel temperature	T_b	°C
Melting point of the polymer	T_m	°C
Viscosity in the melt film	η_f	Pa · s
Velocity of the barrel surface	V_b	cm/s

Velocity component	V_{bx}	cm/s (Fig. 7.10)
Relative velocity	V_J	cm/s
Output of the extruder	G	g/s
Depth of the feed zone	H_1	mm
Width of the screw channel	W	mm
Melt density	ρ_m	g/cm^3
Specific heat of the solid polymer	c_{p_s}	kJ/(kg · K)
Temperature of the solid polymer	T_s	°C
Thermal conductivity of the melt	λ_m	W/(m · K)
Heat of fusion of the polymer	i_m	kJ/kg

Numerical examples illustrating the calculation of melting rate and melting parameter according to Eqns. (7.2.15) and (7.2.16) have been given by RAO in his book [1].

7.2.2.3 Melting Profile

The melting profile gives the amount of unmelted polymer as a function of screw length (Fig. 7.11), and is the basis for calculating the melt temperature and pressure along the screw. It shows whether the polymer at the end of the screw is fully melted. The plasticating and mixing capacity of a screw can be improved by incorporating mixing and shearing devices into the screw. The melting profile enables to judge the suitable positioning of these devices in the screw [4]. Methods of calculating melting profile, melt temperature and melt pressure are given in the book by RAO [4].

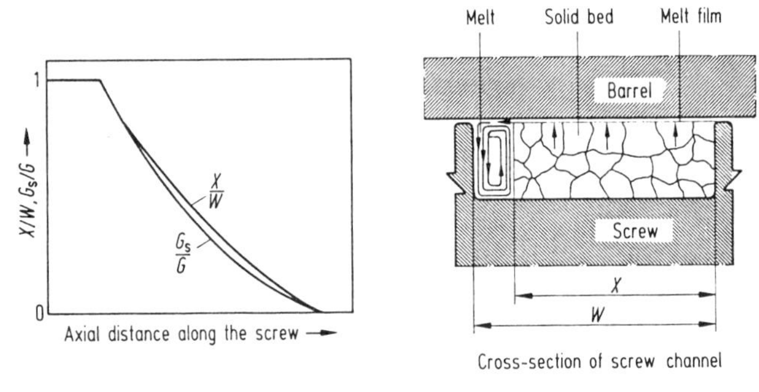

Fig. 7.11: Solid bed and melting profiles X/W and G_S/G [2]
G: total mass flow rate; G_S: mass flow rate of solids

7.2.2.4 Screw Power

The screw power consists of the power dissipated as viscous heat in the channel and flight clearance and the power required to raise the pressure of the melt. The total power Z_N is therefore for a melt filled zone [5]

$$Z_N = Z_c + Z_{FLT} + Z_{\Delta p} \qquad (7.2.17)$$

where Z_c = power dissipated in the screw channel

Z_{FLT} = power dissipated in the flight clearance

$Z_{\Delta p}$ = power required to raise the pressure of the melt

The power $Z_{\Delta p}$ is small in comparison with the sum $Z_c + Z_{FLT}$, and can be neglected. In general the power due to dissipation $\dot{E}_d$ can be expressed for a Newtonian fluid as [1]

$$\dot{E}_d = \eta\, \dot{\gamma}^2 \qquad (7.2.18)$$

Using the respective shear rates for the screw channel (Eq. (7.2.9)) and flight clearance the equation for the screw power can be derived [1].

The power dissipated in the screw channel Z_c is given by [1]

$$Z_c = \frac{\nu \cdot \pi^2 \cdot D_{FLT}^2 \cdot N^2 \cdot W \cdot \eta_c \cdot \Delta L\,(F_x\ \cos^2\phi + 4\sin^2\phi)}{36 \cdot 10^{14} \cdot \sin\phi \cdot H} \qquad (7.2.19)$$

The power dissipated in the flight clearance can be calculated from [1]

$$Z_{FLT} = \frac{\nu \cdot \pi^2 \cdot D_{FLT}^2 \cdot N^2 \cdot W_{FLT} \cdot \Delta L \cdot \eta_{FLT}}{36 \cdot 10_{14} \cdot \delta_{FLT} \cdot \sin\phi} \qquad (7.2.20)$$

The power required to raise the pressure of the melt $Z_{\Delta p}$ can be written as

$$Z_{\Delta p} = 100 \cdot \dot{Q}_p \cdot \Delta p \qquad (7.2.21)$$

The flight diameter D_{FLT} is obtained from

$$D_{FLT} = D_b - 2\delta_{FLT} \qquad (7.2.22)$$

and the channel width W

$$W = \frac{s}{ny} \cos\phi - W_{FLT} \qquad (7.2.23)$$

The symbols and units used in the equations above are given in the following example:

Example:

For the following conditions the screw power is to be determined:

Resin: PE-LD

Operating conditions:

Screw speed N = 80 rpm

melt temperature T = 200 °C

die pressure Δp = 300 bar

Geometry of the metering zone:

D_b = 60 mm; H = 3 mm; e = 6 mm; s = 60 mm; δ_{FLT} = 0.1 mm;

ΔL = 600 mm; ν = 1

Solution:

Power Z_c in the screw channel: D_{FLT} = 59.8 mm from Eq. (7.2.22)

Viscosity of the melt in the screw channel η_c at the average sheer rate 83.8 s^{-1} according to the Eq. (7.2.9) and T = 200 °C from the viscosity function of resin

$\eta_c = 1406.34$ Pa · s

Channel width W = 51.46 from Eq. (7.2.23)

Number of flights $\nu = 1$

Length of the metering zone $\Delta L = 600$ mm

Factor F_x:

$F_x = 1$ for H/W = 3/51.46 = 0.058 from Eq. (7.2.14)

Power in the screw channel Z_c from Eq. (7.2.19)

$$Z_C = \frac{1 \cdot \pi^2 \cdot 59.82 \cdot 802 \cdot 51.46 \cdot 1406.34 \cdot 600 \cdot (1 \cdot \cos^2 17.66° + 4 \cdot \sin^2 17.66°)}{36 \cdot 10^{14} \cdot \ 3 \cdot \sin 17.66°} = 3.85 \text{kW}$$

Power in the flight clearance Z_{FLT}:

Flight width W_{FLT}(Fig. 7.8)

$W_{FLT} = e \cos \Phi = 6 \cdot \cos 17.66° = 5.7$ mm

Viscosity at the average shear rate in the flight clearance

$$\dot{\gamma}_{FLT} = \frac{\pi D_b \cdot N}{60 \cdot \delta_{FLT}} = 2513.3 \text{ s}^{-1}$$

and T = 200 °C from the viscosity function of the resin

$\eta_{FLT} = 219.7$ Pa · s

Power in the flight clearance Z_{FLT} from Eq. (7.2.20):

$$Z_{FLT} = \frac{1 \cdot \pi^2 \cdot 59.8^2 \cdot 80^2 \cdot 600 \cdot 5.7 \cdot 219.7}{36 \cdot 10^{14} \cdot 0.1 \cdot 0.303} = 1.56 \text{ kW}$$

Power to raise the melt pressure $Z_{\Delta p}$:

Pressure flow $\dot{Q}_p$:

$\dot{Q}_p$ from the example in Section 7.2.2.1

$\dot{Q}_p = 1.249 \cdot 10^{-6}$ m³/s

Die pressure Δp:

$\Delta p = 300$ bar

$Z_{\Delta p}$ from Eq. (7.2.20):

$Z_{\Delta p} = 100 \cdot 1.249 \cdot 10^{-6}$ ^· $300 = 0.0375$ kW

Hence the power $Z_{\Delta p}$ is negligible in comparison with the sum $Z_c + Z_{FLT}$.

With increasing flight clearance, as it might occur due to wear in extruders the melt conveying capacity is reduced, since the leakage flow over the flight increases. To compensate for this reduction the screw speed has to be increased, if the output were to remain the same. This leads to an increase in the power Z_N as shown in Table 7.4B.

The power in the flight clearance Z_{FLT} does mot vary much as the opposing effects of viscosity and flight clearance nearly balance each other.

Table 7.4B: Effect of Flight Clearance on Screw Power for a PE-LD

δ_{FLT} mm	Z_{FLT} kW	Z_c kW	Z_N kW	N rpm
0.1	0.711	2.73	3.44	80
0.15	0.71	2.83	3.54	81.5
0.2	0.65	2.93	3.58	83
0.25	0.626	3.0	3.63	84
0.3	0.6	3.15	3.75	86

7.2.2.5 Melt Temperature and Melt Pressure

Melt Temperature

The exact calculation of melt temperature can be done only on an iterative basis as shown in the computer program given in [4]. The temperature rise ΔT in an increment of the screw can be found from [5]

$$\Delta T = (T_{out} - T_m) = \frac{3600 \cdot (Z_c + Z_{FLT} + N_H)}{\dot{m} \cdot c_{pm}} \quad (7.2.24)$$

where Z_c, Z_{FLT} = (see Section 7.2.2.4) (kW)
N_H = heat through the barrel (kW)
$\dot{m}$ = extruder output (kg/h)
T_{out} = temperature of the melt leaving the increment (°C)
T_m = melting point of the polymer (°C)

The average temperature $0.5\,(T_{in} + T_{out})$ can be taken as the stock temperature in the increment, where T_{in} denotes the temperature of the melt entering the increment.

Temperature Fluctuation

Fluctuations of melt temperature in an extruder serve as a measure of the stability of extrusion and extrudate quality.

The temperature variation ΔT can be estimated for a 3-zone screw from the following empirical relationship developed by Squires [19]:

$$\Delta T = \frac{5}{9}\left[\frac{1}{4.31\ N_Q^2 - 0.024}\right] \quad (7.2.25)$$

for $0.11 < N_Q < 0.5$. The parameter N_Q is given by

$$N_Q = 14.7 \cdot 10^{-4} \frac{D_b^2}{G} \Sigma \frac{L}{H} \quad (7.2.26)$$

where ΔT = temperature variation °C
D_b = barrel diameter cm
G = output g/s
L = Length of screw zone in diameters
H = depth of the screw zone cm

Example:

D_b = 6 cm;	G = 15 g/s;	L	H	L/H
		9	0.9	10
		3	0.6 (mean value)	3.33
		9	0.3	30

Hence L/H = 43.33
N_Q from Eq. (7.2.26)

$$N_Q = 14.7 \cdot 10^{-4} \cdot \frac{36}{15} \cdot 43.33 = 0.153$$

ΔT from Eq. (7.2.25)

$$\Delta T = \frac{5}{9}\left[\frac{1}{4.31 \cdot 0.153^2 - 0.024}\right] = 7.22°\,C \quad \text{or} \quad \pm 3.61°\,C$$

The constants occuring in the Eqs. (7.2.26) and (7.2.25) depend on the type of polymer used. For screws other than 3-zone screws the geometry term in Eq. (7.2.26) has to be defined in such a way that N_Q correlates well with the measured temperature fluctuations.

Melt Pressure

The melt or stock pressure can be obtained from the pressure flow by means of Eq. (7.2.4) [1], [5]. The more exact calculation of the melt pressure profile in an extruder should consider the effect the ratio of pressure flow to drag flow, the so-called drossel quotient as shown in [5].

Pressure Fluctuation

The effect of pressure fluctuations of the melt on the flow rate can be estimated by the relation [2]

$$\dot{Q}_1 / \dot{Q}_2 = (\Delta p_1 / \Delta p_2)^n \quad (7.2.27)$$

Example:

pressure drop $\Delta p_1 = 200$ bar; $\Delta p_2 = 210$ bar

flow rate $\dot{Q}_1 = 39.3 \ cm^3/s$

power law exponent n for PE-LD = 2.5
Calculate the flow rate variation.

Solution:

$$\dot{Q}_1 / \dot{Q}_2 = 39.7 / \dot{Q}_2 = (200/210)^{2.5} = 0.885$$

$$\dot{Q}_2 = 44.9 \ \text{cm}^3/\text{s}$$

$$\text{flow rate variation} = (\dot{Q}_2 - \dot{Q}_1)/\dot{Q}_1 = \frac{44.9 - 39.7}{39.7} = 13\%$$

$$\text{pressure variation} = \frac{\Delta p_2 - \Delta p_1}{\Delta p_1} = \frac{210 - 200}{200} = 5\%$$

This means that a pressure variation of 5% can cause a flow rate fluctuation of 13%.

The influence of machine related parameters and operating variables on the melting of solid polymer is shown in Figs. 7.12A-L [2].

For given resin the fluctuation of the melt temperature depends mainly on the screw geometry and output rate as shown in Fig. 7.13 [2].

Empirical design data for different screw geometries, resins and processes is given in Figs. 7.14A-E [11].

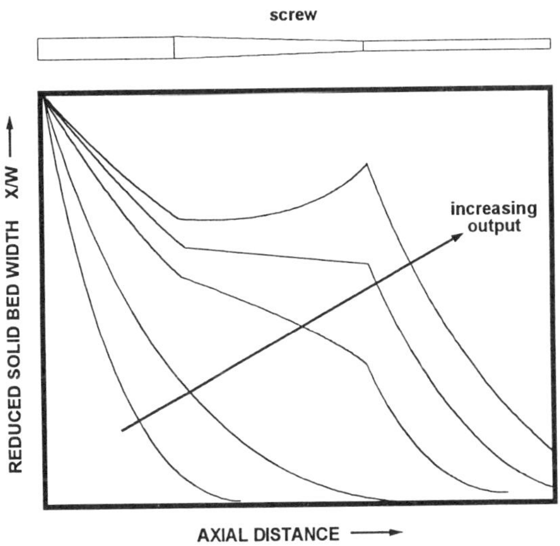

Fig. 7.12A: Influence of increasing output on solids melting in an extruder [2]

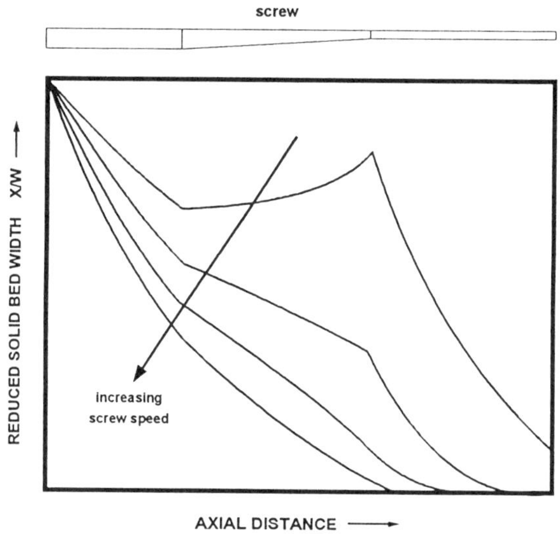

Fig. 7.12B: Influence of increasing screw speed on solids melting in an extruder [2]

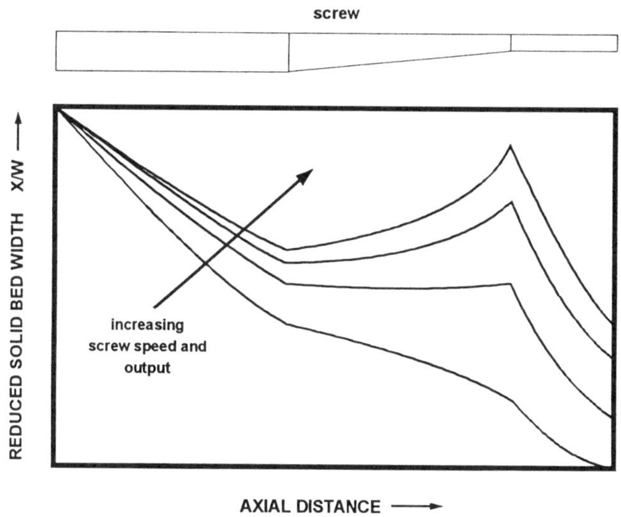

Fig. 7.12C: Influence of increasing screw speed and output on solids melting in an extruder [2]

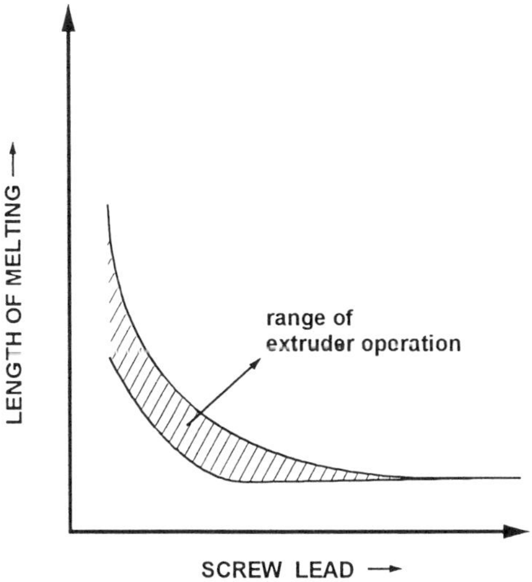

Fig. 7.12D: Relationship between length of melting and screw lead in an extruder [2]

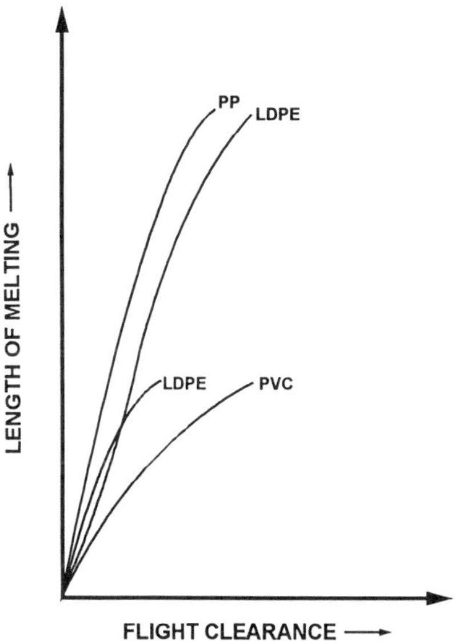

Fig. 7.12E: Relationship between flight clearance and length of melting for different resins in an extruder [2]

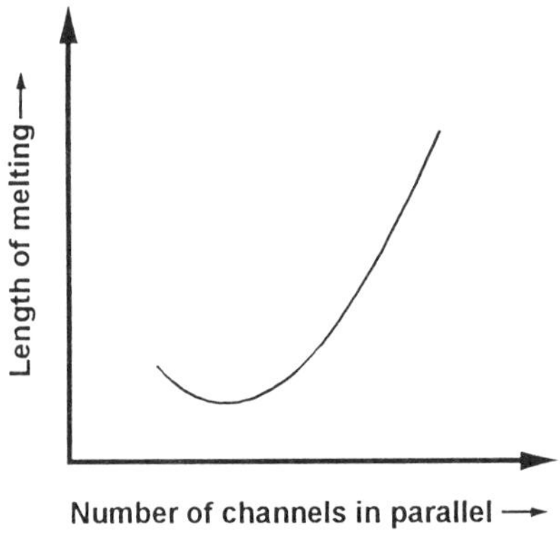

Fig. 7.12F: Relationship between length of melting and number of channels in an extruder [2]

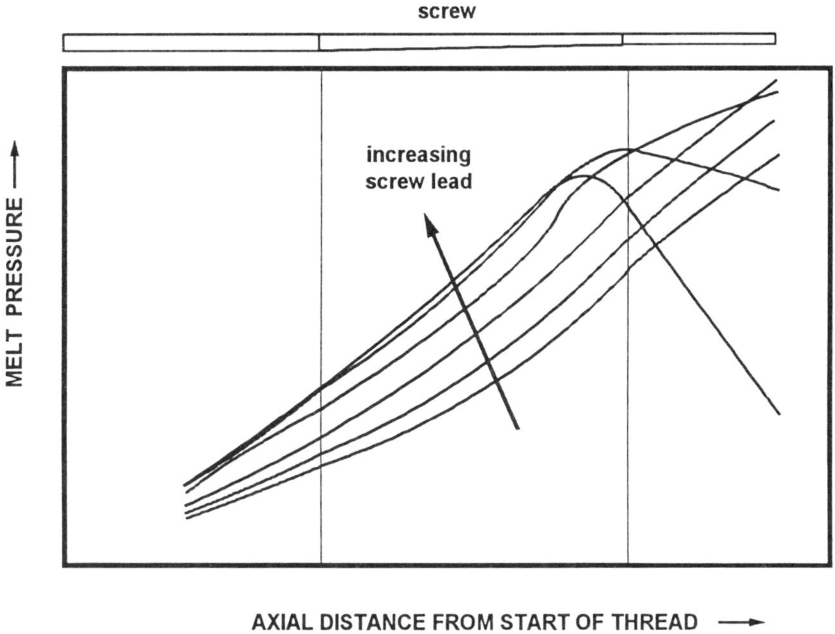

Fig. 7.12G: Relationship between melt pressure and screw length at increasing screw speed[2]

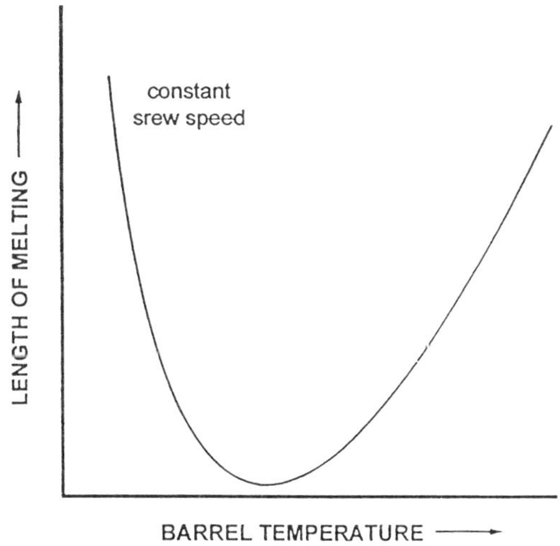

Fig. 7.12H: Relationship between length of melting and barrel temperature at constant screw speed [2]

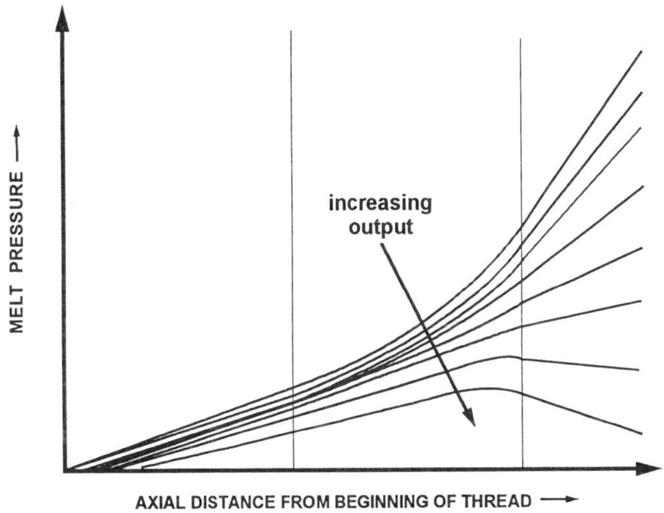

Fig. 7.12I: Relationship between melt pressure and screw length at constant screw speed and increasing output [2]

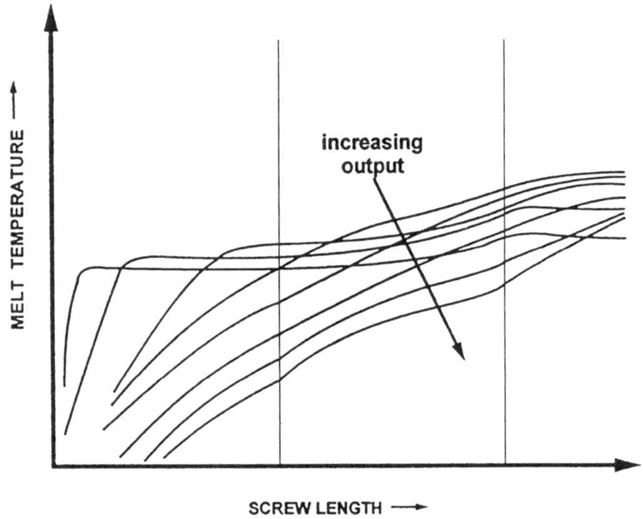

Fig. 7.12J: Relationship between melt temperature and screw length at increasing output [2]

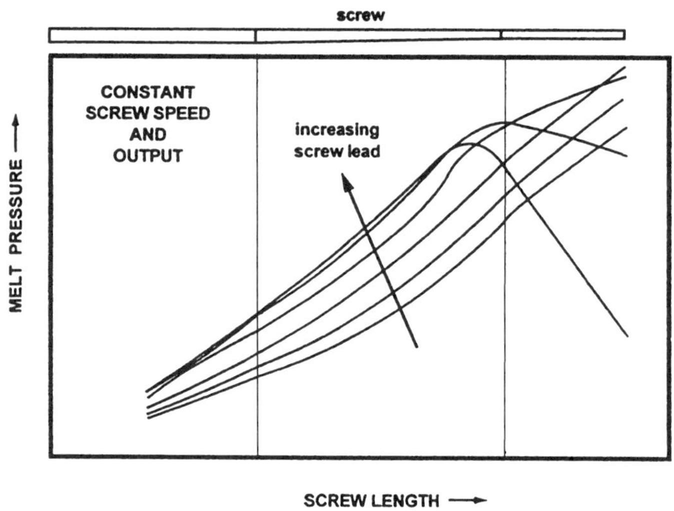

Fig. 7.12K: Relationship between melt pressure and screw length at constant output and screw speed and increasing screw lead [2]

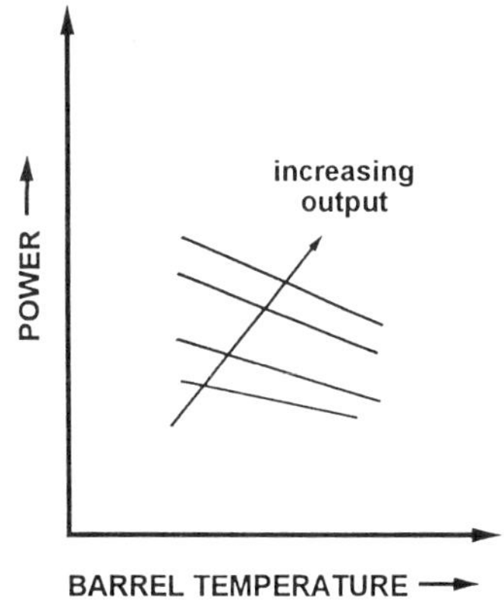

Fig. 7.12L: Relationship between power and barrel temperature at increasing output [2]

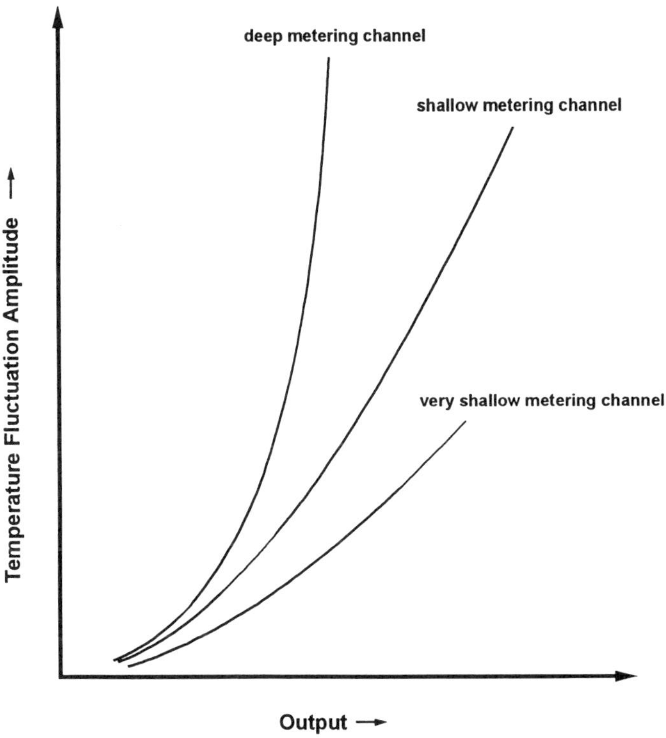

Fig. 7.13: Relationship between melt temperature fluctuation, screw geometry and output [2]

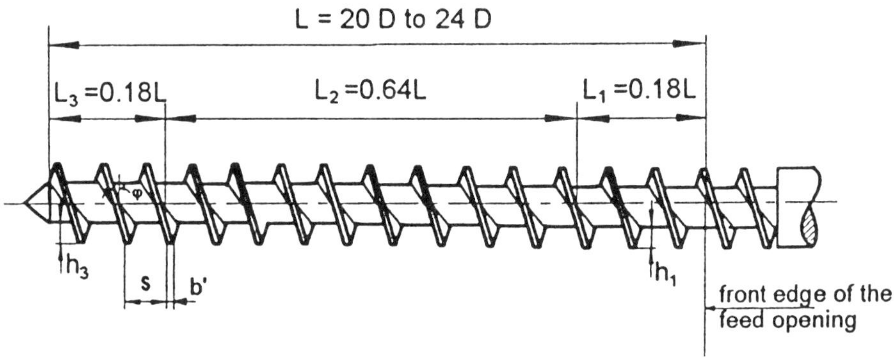

Fig. 7.14A: Empirical screw design data [11]

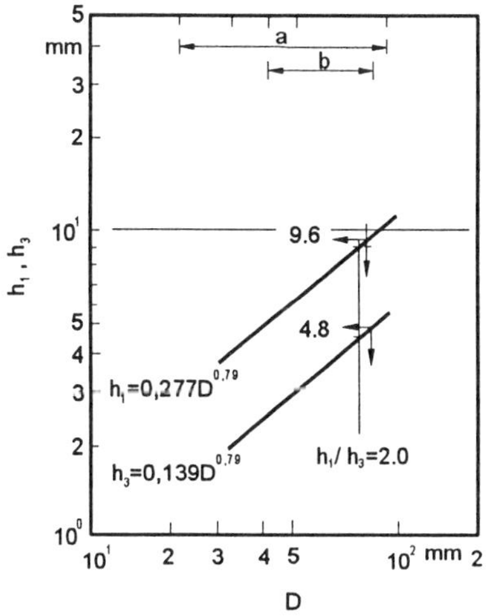

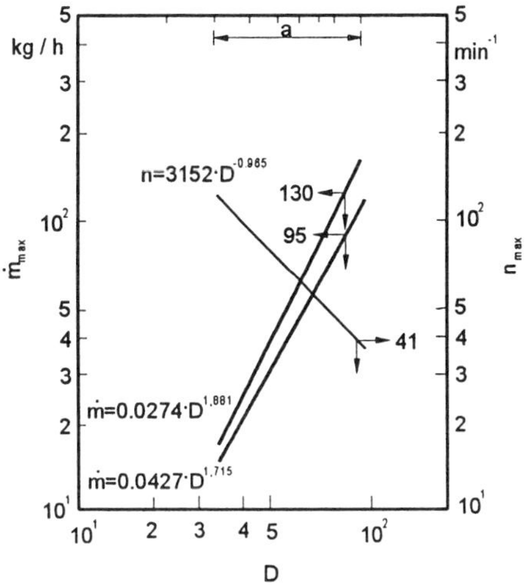

Fig. 7.14B: Empirical screw design data [11]

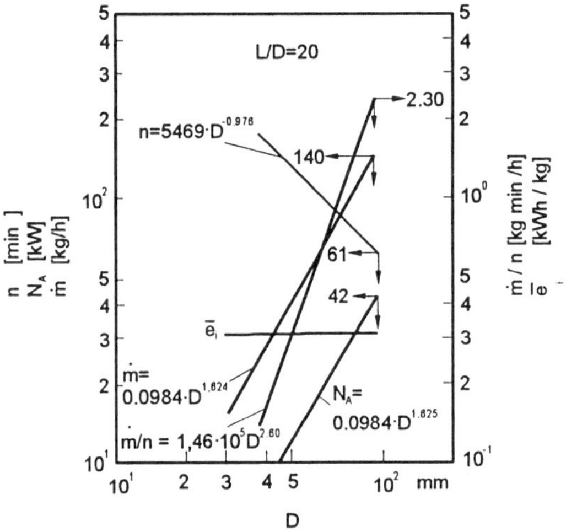

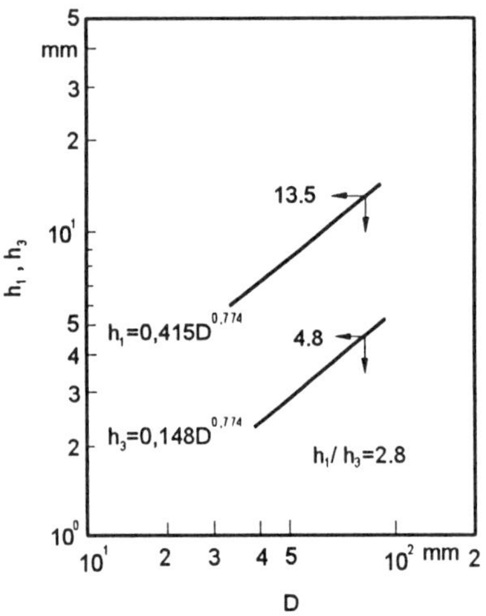

Fig. 7.14C: Empirical screw design data [11]

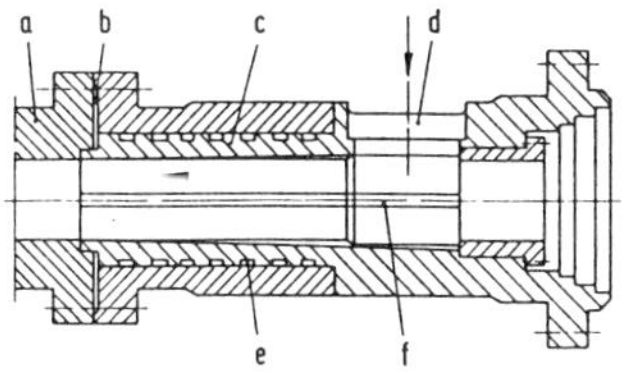

High efficiency feed zone
a extruder barrel, *b* thermal break,
c grooved sleeve, *d* feed opening,
e cooling spiral, *f* grooves

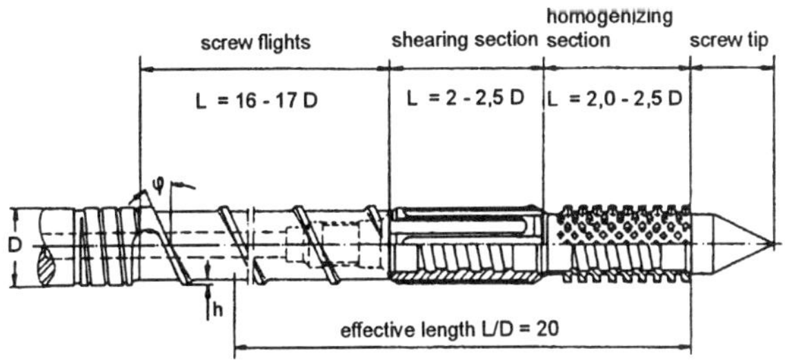

Screw built from standard modules for grooved blow molding extruders

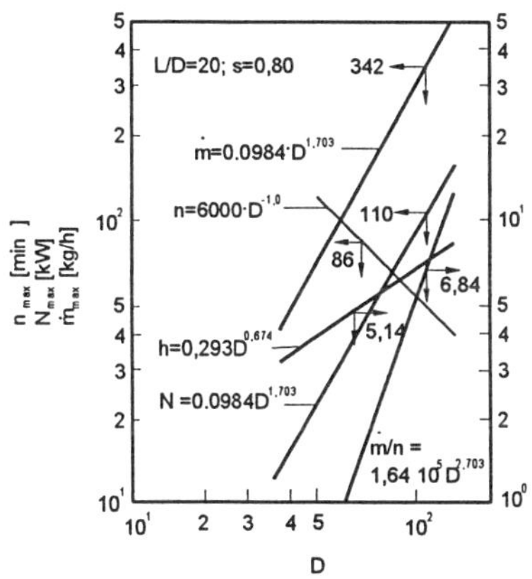

Design data for a series of grooved extruders

Fig. 7.14D: Empirical screw design data [11]

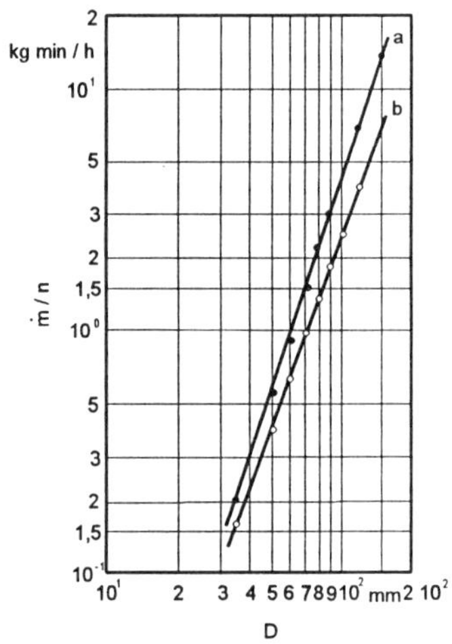

Comparison of specific throughput rates $\dot{m}/n$ of conventional and grooved extruders (material: HDPE, Lupolen 5021 D) a grooved extruder $\dot{m}/n = 6.6 \cdot 10^{-6} \cdot D^{2.9}$, b conventional extruder $\dot{m}/n = 3.1 \cdot 10^{-6} \cdot D^{2.7}$

Fig. 7.14E: Empirical screw design data [11]

7.3 Extrusion Dies

Extrusion dies can be designed by calculating shear rate, die pressure and residence time of the melt as functions of the flow path of melt in the die [6]. Of these quantities the die pressure is most important as the desired throughput cannot be attained if the die pressure does not match with the melt pressure. The interaction between screw and die is shown in Fig. 7.15.

Common shapes of flow channels occurring in extrusion dies are shown in Fig. 7.16. Detailed treatment of die design is presented in the book [1] and in [17]. Following areas of application of extrusion dies serve as examples to illustrate the relationship between die geometry and processing parameters:

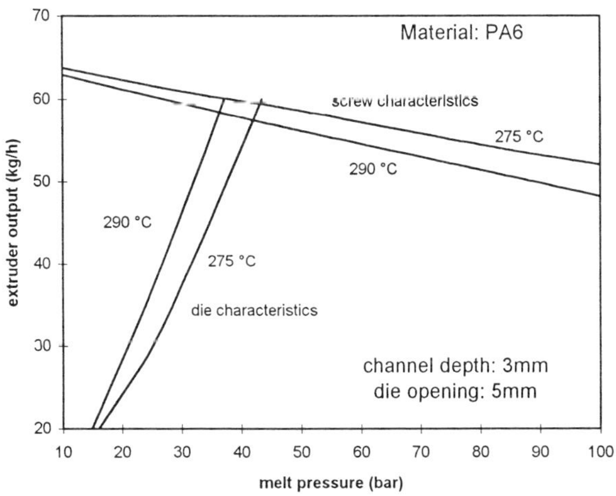

Fig. 7.15A: Effect of screw and die temperature

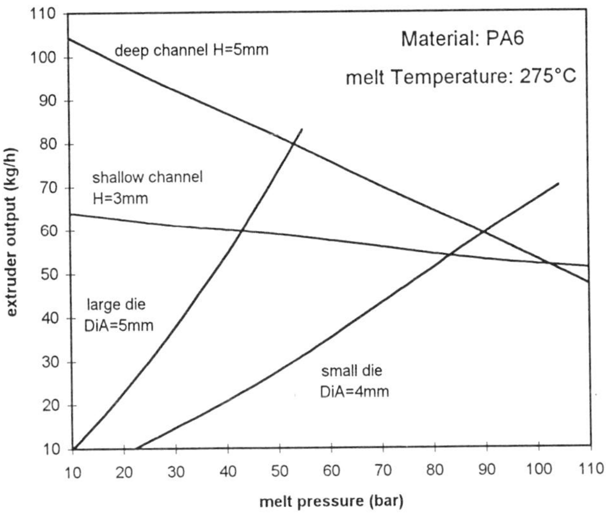

Fig. 7.15B: Effect of channel depth and die opening

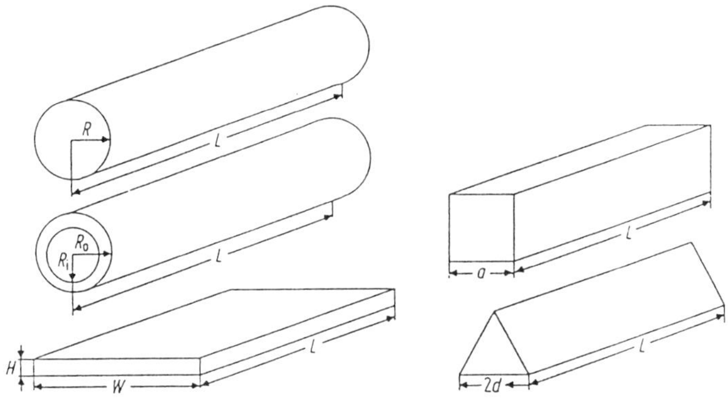

Fig. 7.16: Common shapes of flow channels in extrusion dies [1]

7.3.1 Pipe Extrusion

The spider die shown in Fig. 7.17 is employed for making tubes and pipes and also for extruding a parison required to make a blow-molded article. It is also used in blown film processes.

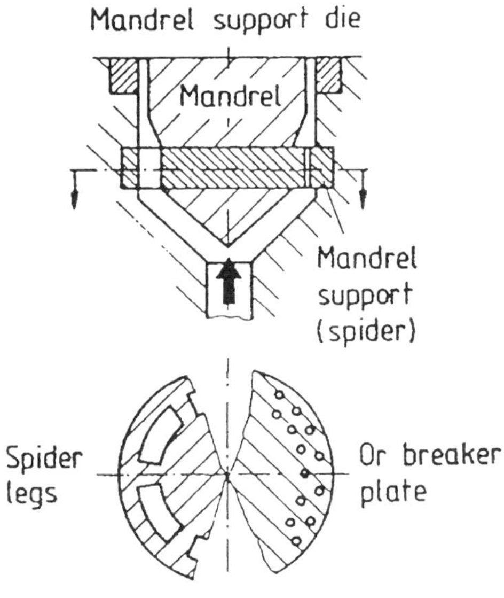

Fig. 7.17: Mandrel support die with spider or break plate [17]

For a circular channel the shear rate is calculated from Eq. (3.2.1). For an annulus, which represents the pipe cross-section it is given by

$$\dot{\gamma}_{annulus} = \frac{6\,\dot{Q}}{\pi\,(R_o + R_i)\,(R_o - R_i)^2} \tag{7.3.1}$$

The pressure drop is obtained from [7]

$$\Delta p = \frac{\dot{Q}^{\frac{1}{n}}}{K^{\frac{1}{n}} \cdot G} \tag{7.3.2}$$

where G, the die constant follows from

$$G_{circle} = \left(\frac{\pi}{4}\right)^{\frac{1}{n}} \cdot \frac{R^{\frac{1}{n}+1}}{2\,L} \tag{7.3.3}$$

$$G_{slit} = \left(\frac{W}{6}\right)^{\frac{1}{n}} \cdot \frac{H^{\frac{2}{n}+1}}{2\,L} \tag{7.3.4}$$

for $W/H \geq 20$. For $W/H < 20$, G_{slit} has to be multiplied by the correction factor F_p given in Fig. 7.18 as a function of H/W. The die constant $G_{annulus}$ is calculated from

$$H = R_o - R_i \tag{7.3.5}$$

$$W = \pi\,(R_o + R_i) \tag{7.3.6}$$

$G_{annulus}$ follows then from

$$G_{annulus} = \left(\frac{\pi}{6}\right)^{\frac{1}{n}} \cdot \frac{(R_o + R_i)^{\frac{1}{n}} \cdot (R_o - R_i)^{\frac{2}{n}+1}}{2\,L} \tag{7.3.7}$$

for $(R_o + R_i) / (R_o - R_i) \geq 37$.
For smaller values of this ratio $G_{annulus}$ has to be multiplied by F_p given in Fig. 7.18. The height H and the width W are obtained in this case from Eqns. (7.3.5) and (7.3.6).

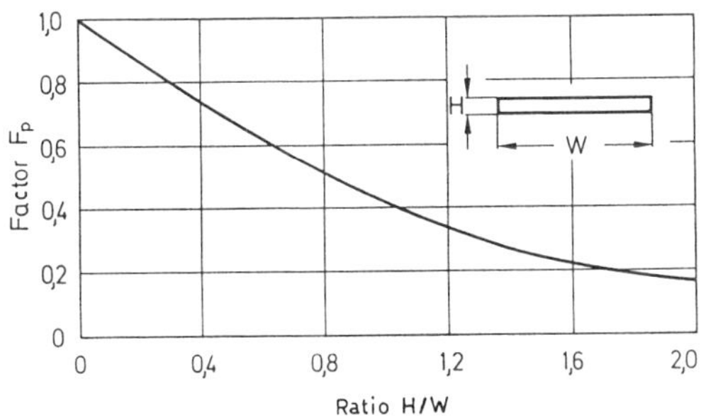

Fig. 7.18 Correction factor F_p as a function of H/W [19]

In Table 7.5 the shear rates and die constants for different channel shapes are summarized.

Table 7.5: Shear Rates and Die Constants for Some Die Channel Shapes (Fig. 7.16) [1]

Channel shape	Shear rate $\dot{\gamma}$ [s^{-1}]	Die constant G
Circle	$4\dot{Q}/\pi\ R^3$	$\left(\frac{\pi}{4}\right)^{\frac{1}{n}} \cdot \frac{R^{\frac{1}{n}+1}}{2\ L}$
Slit	$\frac{6\ \dot{Q}}{W \cdot H^2}$	$\left(\frac{W}{6}\right)^{\frac{1}{n}} \cdot \frac{H^{\frac{2}{n}+1}}{2\ L}$
Annulus	$\frac{6\dot{Q}}{\pi(R_o+R_i)\ (R_o-R_i)^2}$	$\left(\frac{\pi}{6}\right)^{\frac{1}{n}} \cdot \frac{(R_o+R_i)^{\frac{1}{n}} \cdot (R_o-R_i)^{\frac{2}{n}+1}}{2\ L}$
Triangle	$\frac{10}{3} \cdot \frac{\dot{Q}}{d^3}$	$\frac{1}{\sqrt{3}} \cdot \left(\frac{3}{10}\right)^{\frac{1}{n}} \cdot \frac{d^{\frac{3}{n}+1}}{2\ L}$
Square	$\frac{3}{0.42} \cdot \frac{\dot{Q}}{a^3}$	$\frac{1}{2}\left(\frac{0.42}{3}\right)^{\frac{1}{n}} \cdot \frac{a^{\frac{3}{n}+1}}{2\ L}$

General Cross Section

By means of the substitute radius defined by Schenkel [18] the pressure drop in cross sections other than the ones treated above can be calculated. The substitute radius is expressed by [18]

$$R_{rh} = \left[\frac{\left(\frac{2n+1}{\pi}\right) \cdot A^{n+2}}{B^{n+1}} \right]^{\frac{1}{n+3}} \tag{7.3.8}$$

where R_{rh} = substitute radius
A = cross-sectional area
B = circumference

On the basis of the substitute radius the pressure drop in the channel is calculated as in the case of the circular channel [1].

Another method of calculating the pressure drop in channels of varied cross section is presented in Fig. 7.19 [17], [19]. The correction factor F_p and the flow coefficient f_p have the same values in comparable ranges of height to width of the channel.

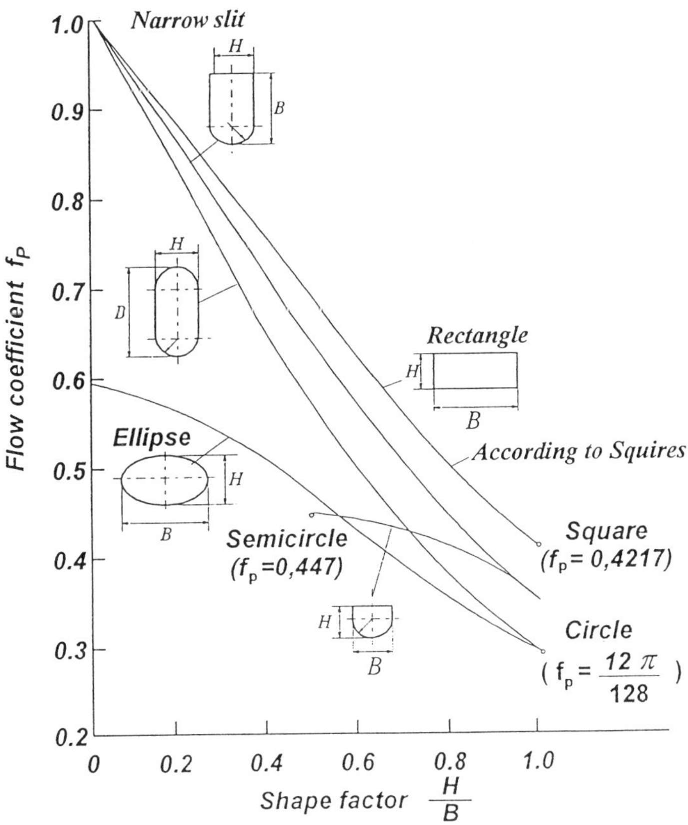

Fig. 7.19: Flow coefficient f_p as a function of shape factor H/B [17]

The symbols and units used in the equations above are explained in the following example:

Example:

It is required to calculate the pressure drop Δp of a PE-LD melt at 200 °C flowing through an annular channel with an outside radius $R_o = 40$ mm, an inside radius $R_i = 39$ mm and length L = 100 mm at a mass flow rate $\dot{m} = 10\ \text{g/s}$.

Solution:

Volume flow rate $$\dot{Q} = \frac{\dot{m}}{\rho_m} = \frac{10}{0.7} = 14.29\ \frac{\text{cm}^3}{\text{s}}$$

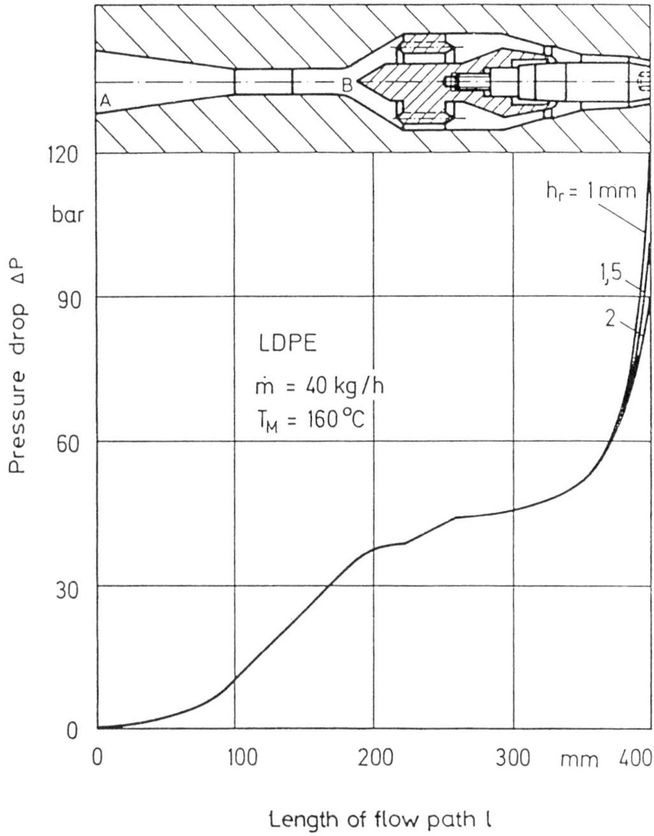

Fig. 7.20A: Melt pressure in a spider die [6]

or $\dot{Q} = 1.429 \cdot 10^{-5}\ m^3/s$

where ρ_m = melt density = 0.7 g/cm^3

Shear rate from Eq. (7.3.1)

$$\dot{\gamma} = \frac{6\dot{Q}}{\pi\ (R_o + R_i)\ (R_o - R_i)^2} = \frac{6 \cdot 1.429 \cdot 10^{-5}}{\pi\ (0.04 + 0.039)\ (0.001)^2} = 354.47\, s^{-1}$$

The given resin-dependent values which can also be calculated from flow curves (see Section 3.2.7.2) are

$n = 2.907$

$\eta = 579.43$ Pa · s

$K = 1.34 \cdot 10^{-13}$

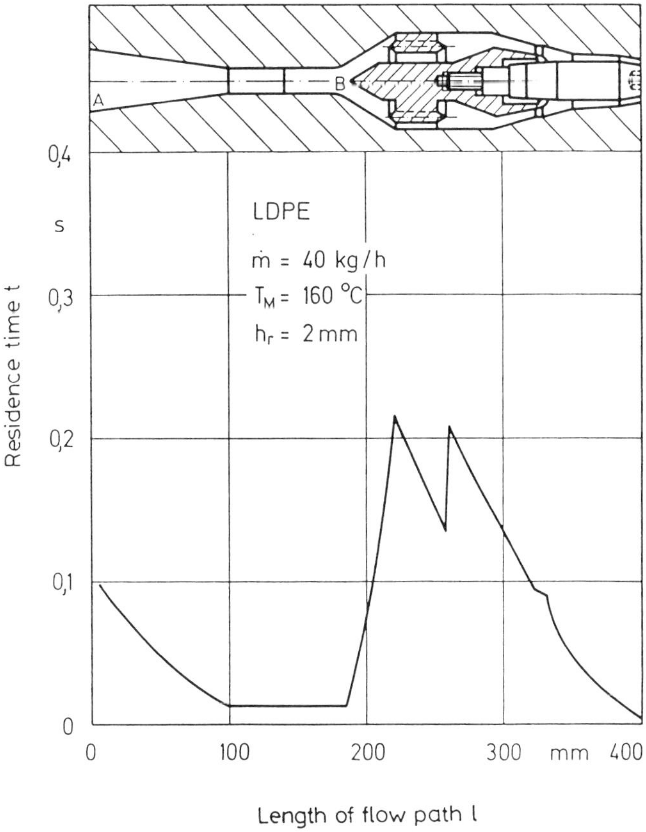

Fig. 7.20B: Residence time in a spider die [6]

The die constant $G_{annulus}$ follows from Eq. (7.3.7)

$$G_{annulus} = \frac{\left(\frac{\pi}{6}\right)^{\frac{1}{2.907}} \cdot (0.04 + 0.39)^{\frac{1}{2907}} \cdot (0.001)^{\frac{2}{2907}+1}}{2 \cdot 0.1}$$

As the ratio $\pi(R_o + R_i)/(R_o - R_i) = 248.19$ is greater than 37, no correction is necessary. Therefore, $G_{annulus} = 1.443 \cdot 10^{-5}$.

The pressure drop Δp is obtained finally from Eq. (7.3.2)

$$\Delta p = 400.26 \text{ bar}$$

The residence time $\bar{t}$ in a channel of length ΔL can be written as

$$\bar{t} = \Delta L / \bar{u}$$

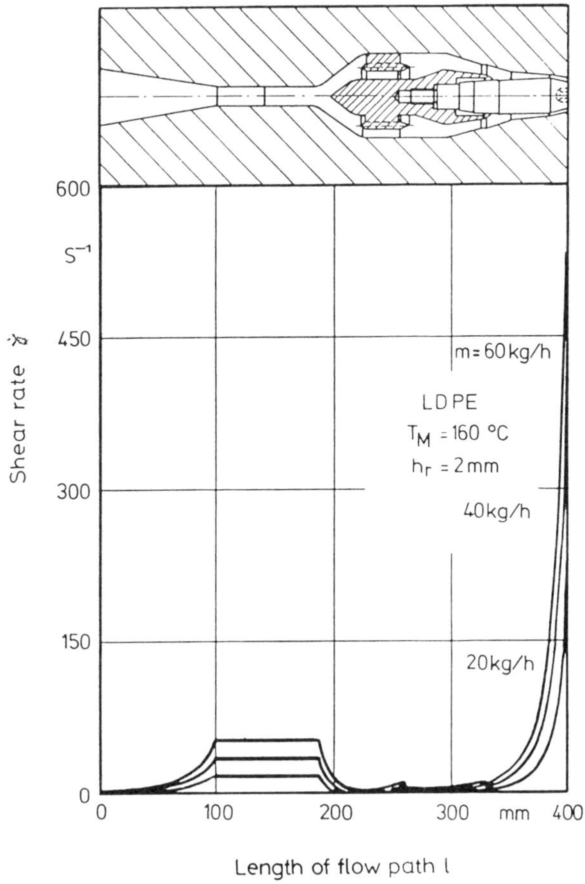

Fig. 7.20C: Shear rate in a spider die [6]

The average velocity $\bar{u}$ can be expressed in terms of shear rate, channel cross section and volume flow rate. The adiabatic temperature increase of the melt in the die can be obtained from

$$\Delta T = \frac{\Delta p}{10 \cdot \rho_m \cdot c_{pm}} \text{ (K)} \tag{7.3.9}$$

where ΔT = temperature rise (K)

Δp = pressure difference (bar)

ρ_m = melt density (g/cm³)

c_{pm} = specific heat of the melt kJ/(kg K)

Using the equations above, the shear rate, residence time and pressure of the melt along a spider die are calculated and shown in Figs. 7.20A-C. As seen from these figures the die gap has a marked influence on the pressure [6].

Drawdown and Haul-Off Rates

Drawdown occurs when the velocity of the haul-off is greater than the velocity of the extrudate at the die exit. This leads to a reduction of the extrudate cross-section. The draw ratio D_R can be expressed as (Fig. 7.21)

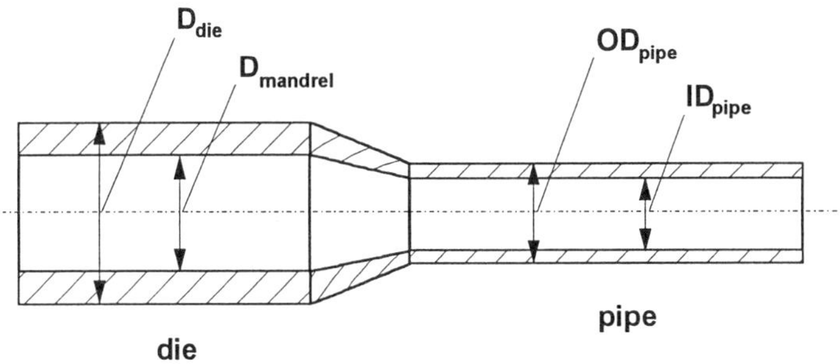

Fig. 7.21: Draw-down in pipe extrusion

$$D_R = \frac{D_{die}^2 - D_{mandrel}^2}{OD_{pipe}^2 - ID_{pipe}^2} \tag{7.3.10}$$

The draw ratio is dependent on the resin and on the haul-off rate. In Fig. 7.22 [8] the ratio of die diameter to pipe diamteter which is referred in practice as the draw ratio is shown as a function of the haul-off rate for a PA resin. For other resins this ratio has to be determined experimentally.

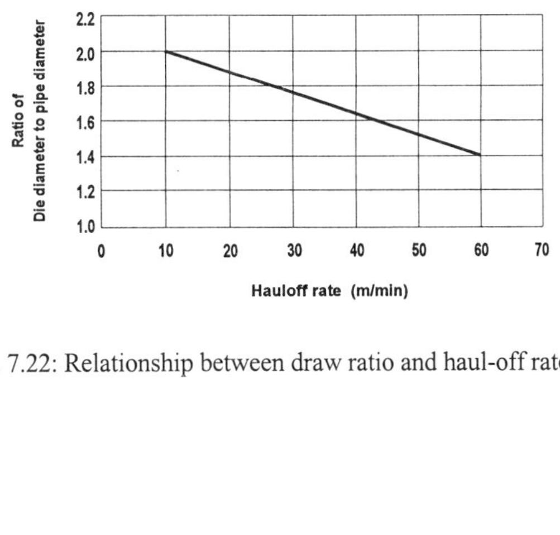

Fig. 7.22: Relationship between draw ratio and haul-off rate [8]

Table 7.6 shows typical data obtained on twin screw extruders for pipes. Design data for pipe dies and sizing units (Fig. 7.23) is given in Table 7.7 [12].

Table: 7.6: Typical Data for Twin-Screw Extruders for Pipes out of Rigid PVC [12]

Screw diameter D mm	Screw length L/D	Screw speed min^{-1}	Screw power kw	specific energy kwh/kg
60/70	18/22	35/50	15/25	0.1/0.14
80/90	18/22	30/40	28/40	0.1/0.14
100/110	18/22	25/38	58/70	0.1/0.14
120/140	18/22	20/34	65/100	0.1/0.14

Table 7.7: Die and Calibration Unit Dimensions Based on Empirical Results [12]

Pipe raw material	Outer diameter of pipe mm	A mm	B mm	S in % of nominal wall thickness	a at pressure 6 to 10 mm	b mm	Haul-off speed m/min
PVC	20	20	20.16	4	30	100/150	20/35
	160	160	161.3	79	150	500/600	2.0/3.5
PE	20	21	21	100	20	150	25/30
	160	168	167.2	100	40/175[1]	40	1.2/2.2
PP	20	21	21	100	20	150	25/30
	160	168	167.2	100	40/175[1]	640	1.0/2.0
					S [mm]		
PA 12[1]	8	14.2	8.6	2.0	25	130	55/60
PA 12[2]	20	28.2	20.85	3.3	25/30[3]	130	12/15
PA 12[2]	22	30.0	23.0	3.3	35/50[3]	130	10/12

[1] wall thickness 1 mm

[2] wall thickness 2 mm

[3] dependent on wall thickness

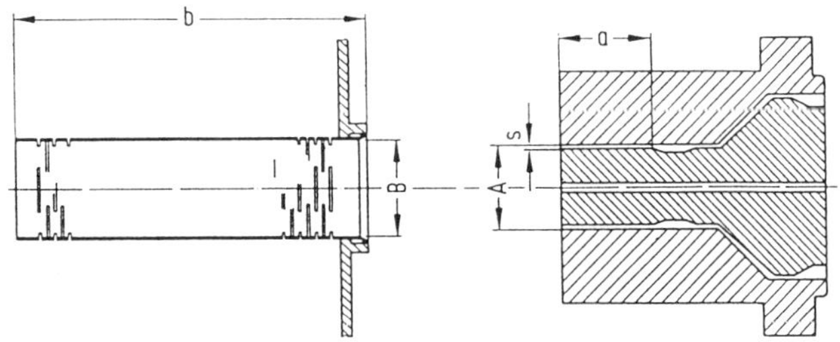

Fig. 7.23. Design data for pipe dies and sizing units (see Table 7.7) [12]

7.3.2 Blown Film

The blow ratio is the ratio between the diameter of blown film to the die lip diameter (Fig. 7.24). The effect of operating variables such as coolant temperature in the case of water-cooled films on some film properties like gloss and haze is shown in Fig. 7.25 [9]. Depending on the material and the type of film blow ratios range from 1.3 : 1 to 6 : 1 [26].

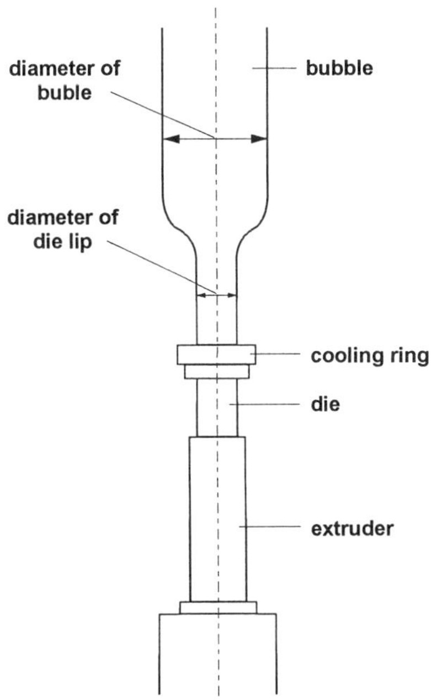

Fig.7.24: Blow-up in blown film

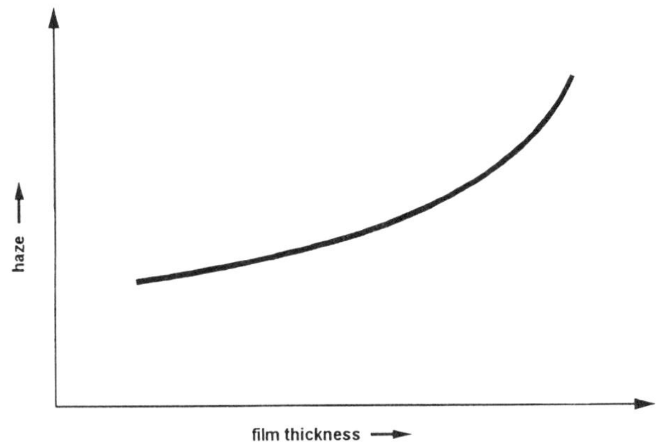

Fig. 7.25A: Effect of film thickness on haze [9]

Extruder throughputs for blown film lines for polyethylene are given in Table 7.8 [13].

Table 7.8: Data for High Performance PE-LD Blown Films Lines [13]

Screw diameter D mm	Maximum throughput kg/h	Extruder power kW
60	200	55
90	400	110
120	550	170
150	900	300

Side fed and spiral mandrel dies are employed in processes, in which the extruder is at an angle, usually 90°, to the direction of the extrudate (Fig. 7.26 [17]). These processes include making annular parisons for blow molding, cable coating and pipe jacketing, to mention a few examples.

In the extrusion of blown film spiral mandrel dies (Fig. 7.26 [17]) are widely used today. The mixing of spiral flow and annular flow of the melt in these dies leads to uniform flow distribution at the die exit, and owing to the lack of spider legs which disturb the melt flow in a spider die there will be no weld lines on the film.

Side fed and spiral dies can also be designed using the equations given in the Section 7.3.1. The principle here is to divide the die into a number of segments and by applying the pressure balance.

$$\Delta p_{\text{spiral flow in the channel}} = \Delta p_{\text{annular flow in the gap}}$$

to each segment calculate the height of the channel required for uniform flow. This can be done conveniently by means of a computer program such as VISPIRAL [20].

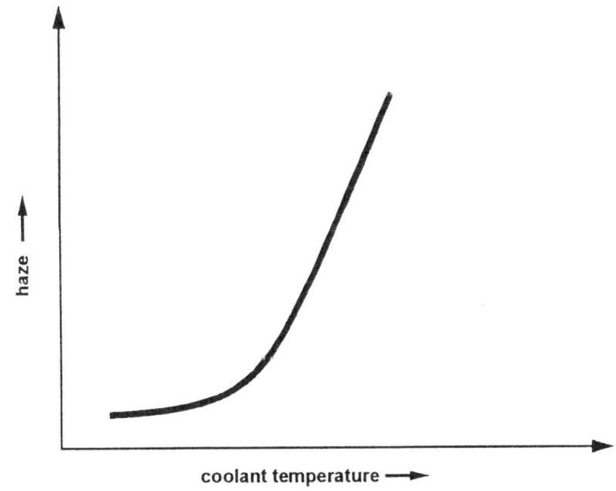

Fig. 7.25B: Effect of coolant temperature on haze [9]

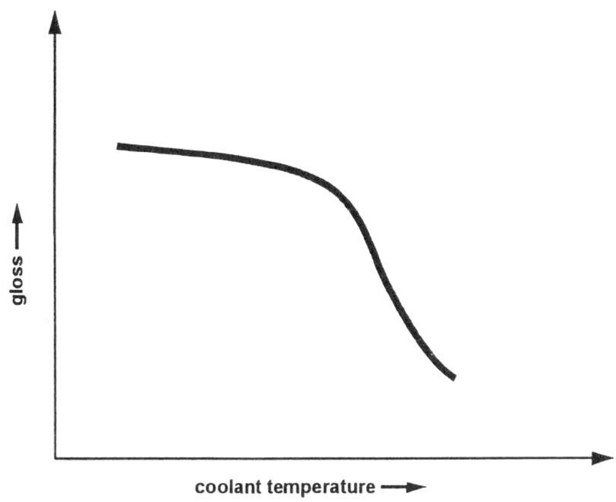

Fig. 7.25C: Effect of coolant temperature on gloss [9]

7.3.3 Sheet extrusion

Dies for sheet and flat film extrusion have a rectangular exit cross-section (Fig. 7.16), where the die width W is usually much larger than the height of the slit H. The shear rate in a rectangular channel is calculated from

$$\dot{\gamma} = \frac{6\,\dot{Q}}{WH^2} \tag{7.3.11}$$

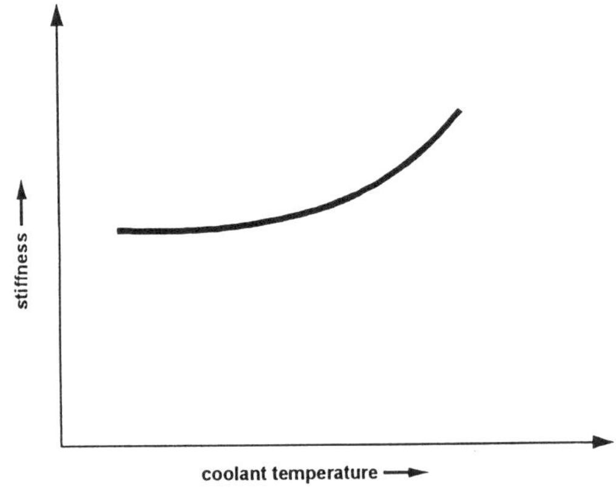

Fig. 7.25D: Effect of coolant temperature on stiffness [9]

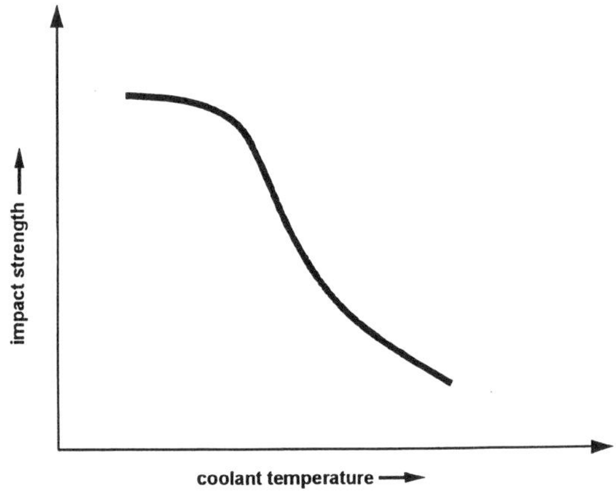

Fig. 7.25E: Effect of coolant temperature on impact strength [9]

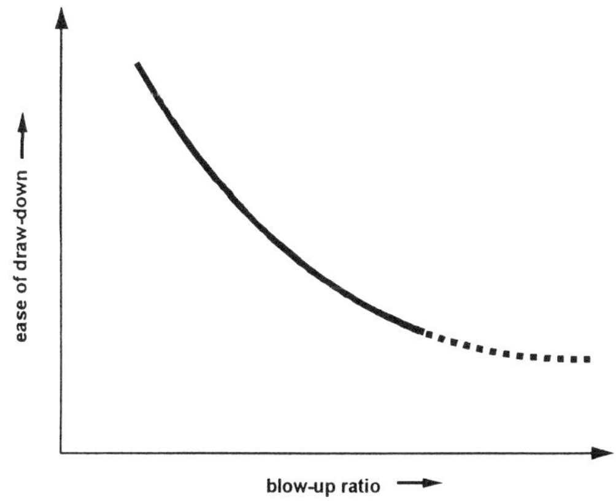

Fig. 7.25F: Effect of blow-up ratio on ease of draw-down [9]

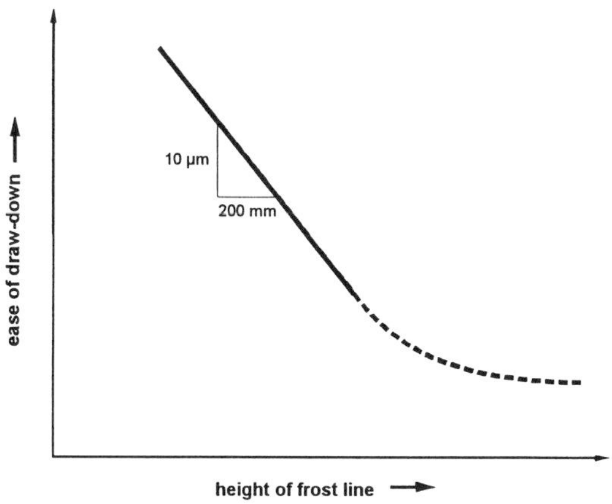

Fig. 7.25G: Effect of frost line on ease of draw-down [9]

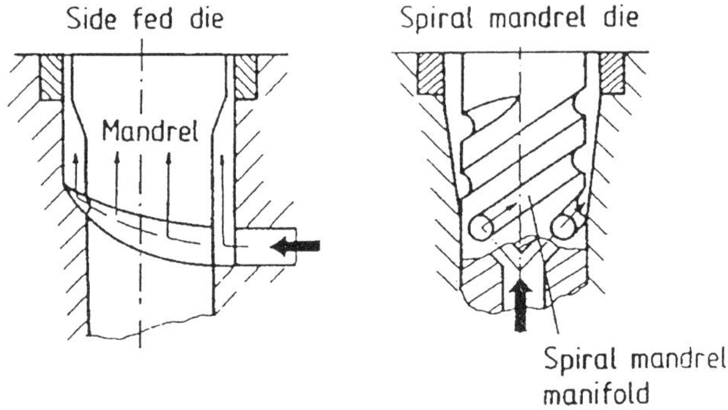

Fig. 7.26: Side-fed spiral mandrel dies [17]

The pressure drop in a sheet die can be obtained from the relation given by Chung and Lohkamp [10]. For a uniform flow distribution the radius R(x) along the manifold (Fig. 7.27) is deduced to be [10]

$$R(x) = R_o \left(1 - \frac{x}{L_M}\right)^{\frac{n_r}{(3 \cdot n_r + 1)}} \tag{7.3.12}$$

where n_r is the reciprocal of the power law exponent n in Eq. (3.2.8).

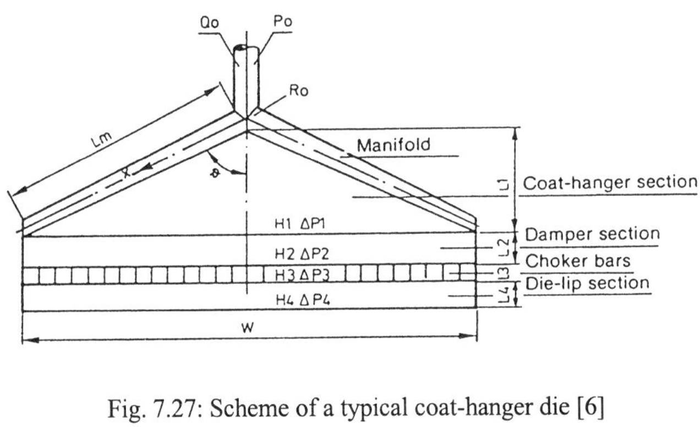

Fig. 7.27: Scheme of a typical coat-hanger die [6]

In Fig. 7.28 the calculated manifold radius R(x) is shown as a function of the distance x along the manifold [6]. The pressure drop in the die lip section is given in Fig. 7.28A as a function of the die gap.

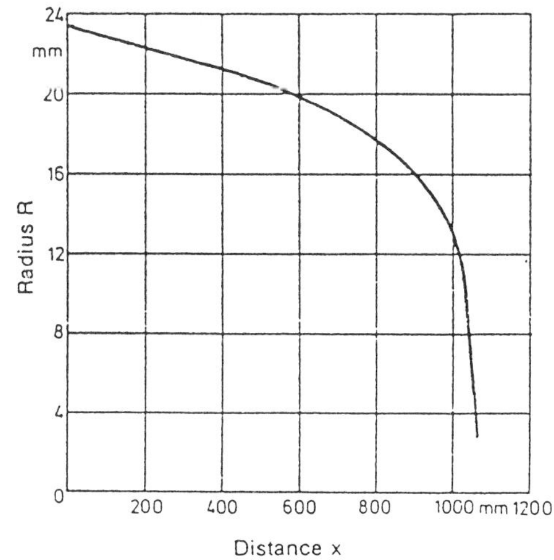

Fig. 7.28: Manifold radius R(x) as a function of the distance x along the manifold [6]

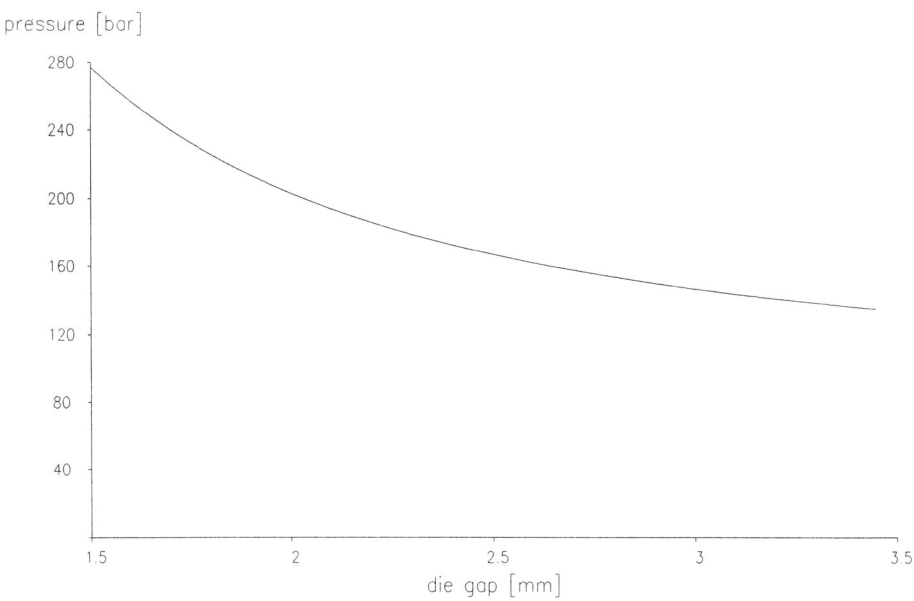

Fig. 7.28A: Pressure drop in a flat die as a function of die gap

7.4 Thermoforming

In thermoforming the material initially in the form of a sheet or film is shaped under vacuum or pressure after it has reached a particular temperature. At this temperature the polymer must have a sufficiently strong viscous component to allow for flow under stress and a significant elastic component to resist flow in order to enable solid shaping. Thus the optimal conditions for thermoforming occur at a temperature corresponding to the material's transition from a solid rubber state to a viscous liquid state [21].

The effect of different factors related to thermoforming is presented in Figs. 7.29A-E [22] [23]. Some data including machine dependent parameters are given in the Tables 7.9 to 7.12.

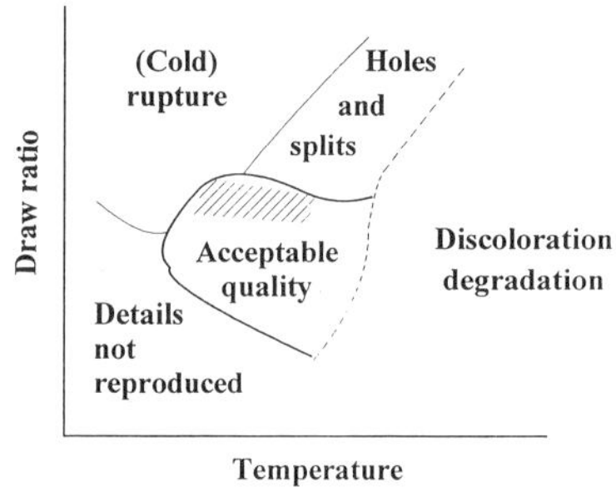

Fig. 7.29A: Parametrical relationship for thermoforming: Draw ratio vs temperature

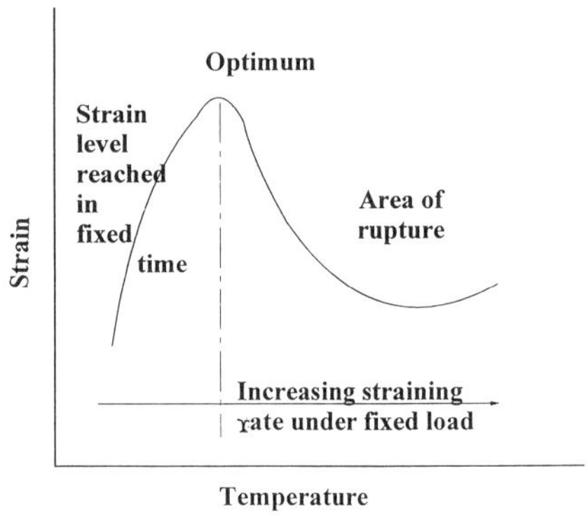

Fig. 7.29B: Parametrical relationship for thermoforming: Strain vs temperature

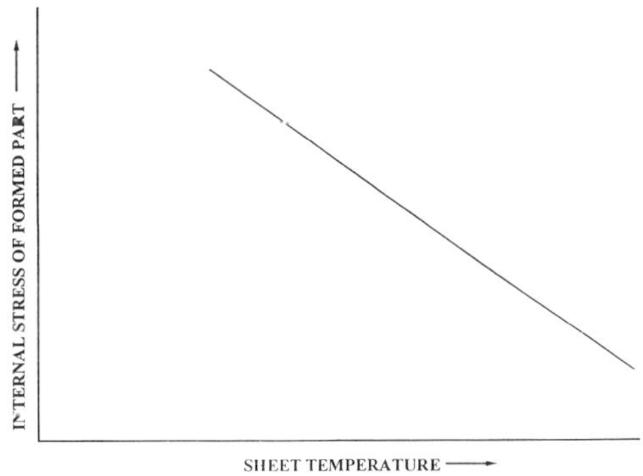

Fig. 7.29C: Parametrical relationship for thermoforming: Internal stress vs. sheet temperature

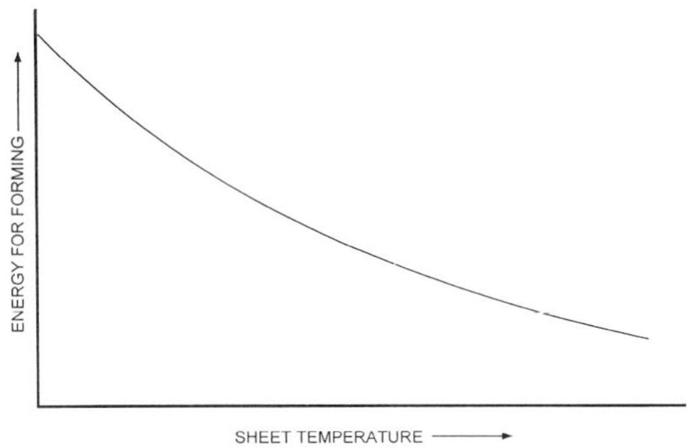

Fig. 7.29D: Parametrical relationship for thermoforming: Energy vs. temperature

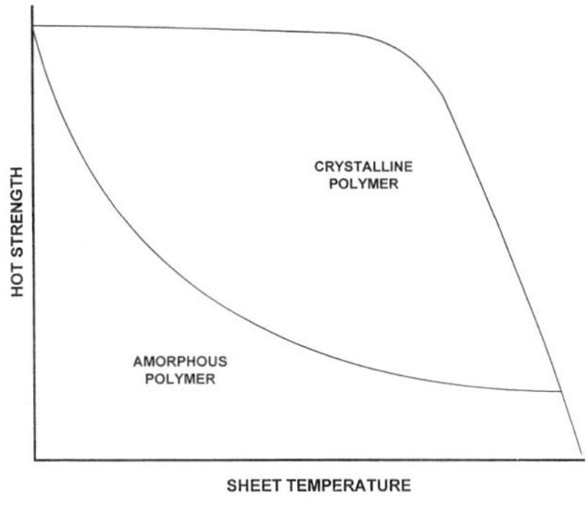

Fig. 7.29E: Parametrical relationship for thermoforming: Hot strength vs. sheet temperature

The production rates for flat and thermoformable films are given in Figs. 7.30A/B [14] and in Table 7.9 [14]. Table 7.10 shows the thermoforming temperature ranges for various thermoplastics [14].

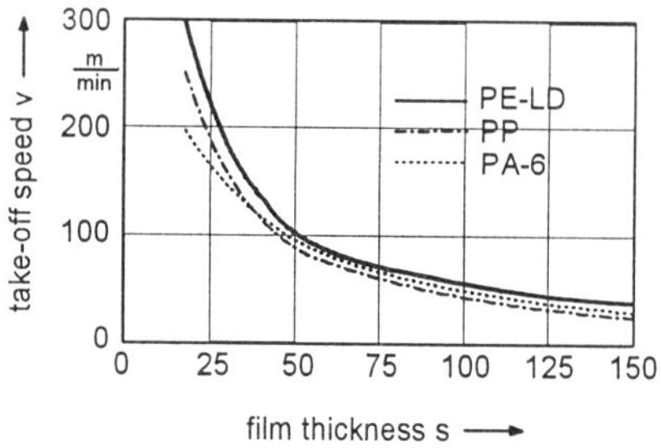

Fig. 7.30A: Production rates for flat films [14]

Table 7.9: Guide to Thermoforming Processing Temperatures [°C] [24]

Polymer	Mold temperature	Lower processing limit	Normal forming heat	Upper limit	Set temperature
PE-HD	71	127	146	166	82
ABS	82	127	163	193	93
PMMA	88	149	177	193	93
PS	85	127	146	182	93
PC	129	168	191	204	138
PVC	60	100	135	149	71
PSU	160	200	246	302	182

Table 7.10: Extruder Outputs for Thermoformable Film Extrusion (Film Thickness Range: 0.4 to 2.0 mm) [14]

Screw diameter D mm	Screw length	Output kg/h			
		PP	SB	ABS	PET
75	30/36 D	180/200	300/320	220/250	120/140
90	30/36 D	260/290	450/500	360/400	180/220
105	30/36 D	320/350	600/650	450/480	240/280
120	30/36 D	480/550	750/850	600/650	320/360
150	30/36 D	650/750	1100/1200	850/900	480/540

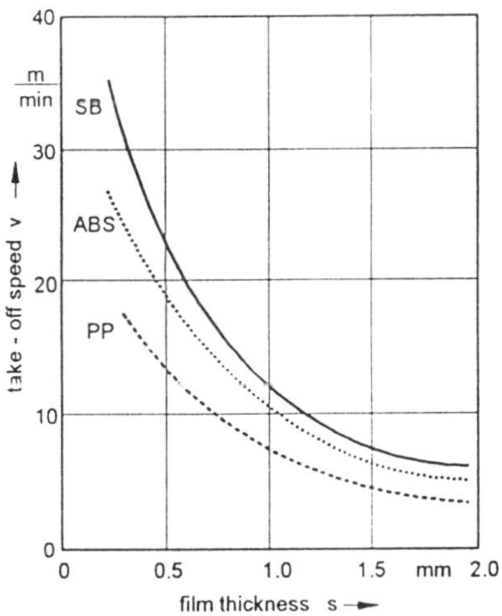

Fig. 7.30B: Production rates for thermoformable films [14]

Table 7.11: Ranges of Melt and Roll Temperature Ranges Used in Thermoformable Film [14]

Material	Melt temperature °C	Chill roll temperature °C
PP	230/260	15/60
SB	210/230	50/90
ABS	220/240	60/100
PET	280/285	15/60

Table 7.12: Shrinkage Guide for Thermoformed Plastics [24]

Polymer	Shrinkage %
PE-LD	1.6/3.0
PE-HD	3.0/3.5
ABS	0.3/0.8
PMMA	0.2/0.8
SAN	0.5/0.6
PC	0.5/0.8
PS	0.3/0.5
PP	1.5/2.2
PVC -U	0.4/0.5
PVC-P	0.8/2.5

7.5 Compounding

In the compounding of polymers such as polyolefins and PVC twin screw extruders find a wide application. Some machine data are given in the tables below.

Table 7.13: Machine Data of Non-Intermeshing Counter-Rotating Twin-Screw Machines for Degassing [25]

Screw diameter D mm	Screw speed n min^{-1}	Drive power N kW	Throughput G kg/h
150	100/180	120/250	670/1250
200	87/150	220/460	1200/2300
250	78/135	370/750	2000/3750
305	70/120	600/1200	3300/6000
380	63/110	970/2000	5400/10000
400	58/100	1500/3000	8300/15000

Table 7.14: Machine Data of Intermeshing, Co-Rotating Twin-Screw Extruders (ZSK) for the Concentration of Melts [25]

Screw diameter D mm	Screw speed n min^{-1}	Drive power N kW	Throughput G kg/h
130	180/300	150/240	850/1500
170	150/250	300/490	1750/3000
240	140/230	680/1100	4000/7000
300	110/180	1200/2000	7000/12000

Table 7.15: Data for the Reifenhaeuser-Bitruder with Intermeshing Counter-Rotating Screws for Compounding and Pelletizing of PVC-U and PVC-P [23]

Extruder	Screw diameter mm	Screw length L/D	Drive power N kW	Heating capacity kW	Output kg/h
BT 80 - 12 G	77	12	25	17	120/230
BT 80 - 16 G	77	16	25	17	200/400
BT 100 - 14 G	98	14	33	31	300/550
BT 100 - 18 G	98	18	37	36	450/750
BT 150 - 17 G	150	17	133	88	650/1200

Coextrusion

In many cases a plastics product made from a single polymer cannot meet the requirements imposed on it. To create layered structures and benefit from the properties of several resins in combination, coextrusion is evidely employed. Examples include multiple layer flat and tubular films, cables with multiple layer insulation, multiple layer blow molded articles and many more.

To achieve these objectives, dies of different designs are used as shown in Fig. 7.30C and in Fig. 7.30D [17]. A multilayer extrudate can be produced by conventional dies when an adapter is used to feed the individual melt streams into the die inlet. They flow together through the die and leave it as a coextrudate (Fig. 7.30C) [17].

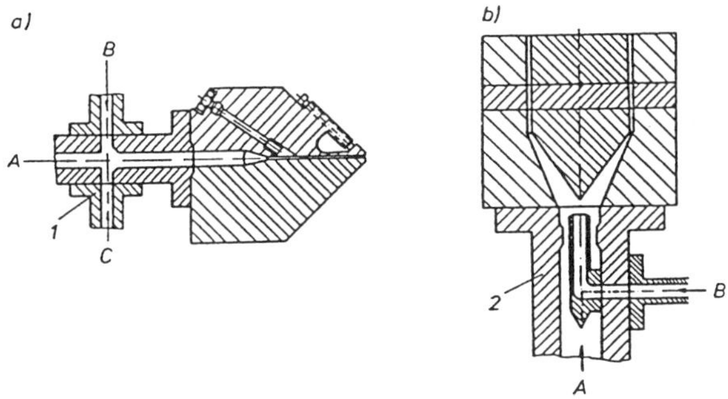

Fig. 7.30C: Adapter dies: a) Flat slit die, b) Blown film die

In the multi-manifold dies (Fig. 7.30D) [17] each melt is first fed separately and distributed into the desired form. These partial streams are then combined just prior to the land area.

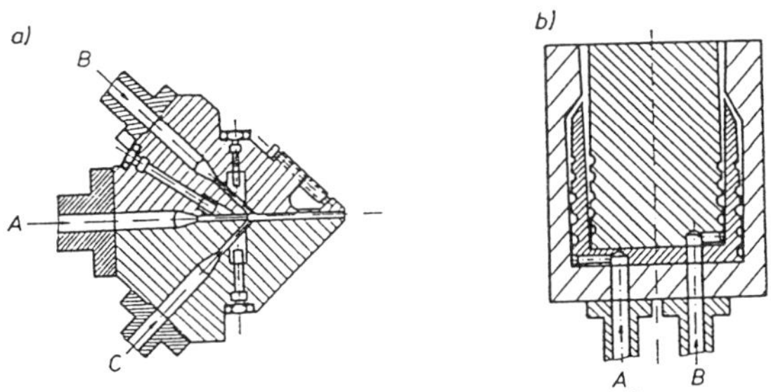

Fig. 7.30D: Multimanifold dies: a) Flat slit die, b) Blown film die [17]

Some advantages and disadvantages of adapter and multi-manifold dies are summarized in Table 7.15A.

Table 7.15A: Comparison Between Adapter (Feedblock) Dies and Multi-Manifold Dies [17]

Type of die	Advantages	Disadvantages
Adapter (Feedblock) dies	Any number of individual layers can be combined.	Polymers used must have almost identical flow behavior and processing temperatures.
Multi-manifold dies	1. Each melt can be adjusted individually, if appropriate adjustment is available. 2. The method of combining the melts inside the die under pressure improves the mutual adhesion of the layers. 3. Materials with different flow behavior (Table 7.16) and melt temperature can be processed	1. Die design is complicated. 2. It is difficult to solve the problem of thermal insulation of the individual channels from each other. 3. For a combination of more than four layers this type of die becomes very complex and costly.

Table 7.16: Examples of Compatibility Between Plastics for Co-Extrusion [24]

	PE-LD	PE-HD	PP	IONOMER	PA	EVA
PE-LD	3	3	2	3	1	3
PE-HD	3	3	2	3	1	3
PP	2	2	3	2	1	3
IONOMER	3	3	2	3	3	3
PA	1	1	1	3	3	1
EVA	3	3	3	3	1	3

1: Layers easy to separate
2: Layers can be separated with moderate effort
3: Layers difficult to separate.

7.6 Extrudate Cooling

The cooling of extrudate is an important operation in downstream extrusion processing, as it sets a limit to the production rate. It is an unsteady heat transfer process, in which mostly conduction and convection determine the rate of cooling. Transient conduction can be described by means of Fourier number and simultaneous conduction and convection by Biot number. The magnitude of heat transfer coefficients which are necessary to estimate the rate of cooling is given in Table 7.17 for different cooling media.

Table 7.17: Heat Transfer Coefficients for Different Types of Cooling

Mode of cooling	Heat transfer coefficient α [W/(m^2 · K)]
Air cooling by natural convection	5/10
Air cooling by forced convection	20/80
Water bath	1000/1800
Water sprays	2000/2500

Example [28]:
A wirc of polyacetal of diameter 3.2 mm is extruded at 190 °C into a water bath at 20 °C. Calculate the length of the water bath to cool the wire from 190 °C to a center-line temperature of 140 °C for the following conditions:

Heat transfer coefficient	α_a	=	1700 W/(m^2 · K)
Thermal diffusivity	$a_{plastic}$	=	10^{-7} m^2/s
Thermal conductivity	$\lambda_{plastic}$	=	0.23 W/(m · K)
Haul-off rate of the wire	V_H	=	0.5 m/s

Solution:

The Biot number $\quad Bi = \dfrac{\alpha_a \cdot R}{\lambda}$

where R = radius of the wire

$$Bi = \frac{1700 \cdot 1.6}{1000 \cdot 0.23} = 11.13 \; ; \quad \frac{1}{Bi} = 0.0846$$

The temperature ratio θ_{T_b}:

$$\theta_{T_b} = \frac{(T_b - T_w)}{(T_a - T_w)} = \frac{140 - 20}{190 - 20} = \frac{120}{170} = 0.706$$

The temperature ratio θ_{T_b} based on the center-line temperature is given in Fig. 7.31 as a function of the Fourier number with the recoprocal of Biot number as parameter [29].

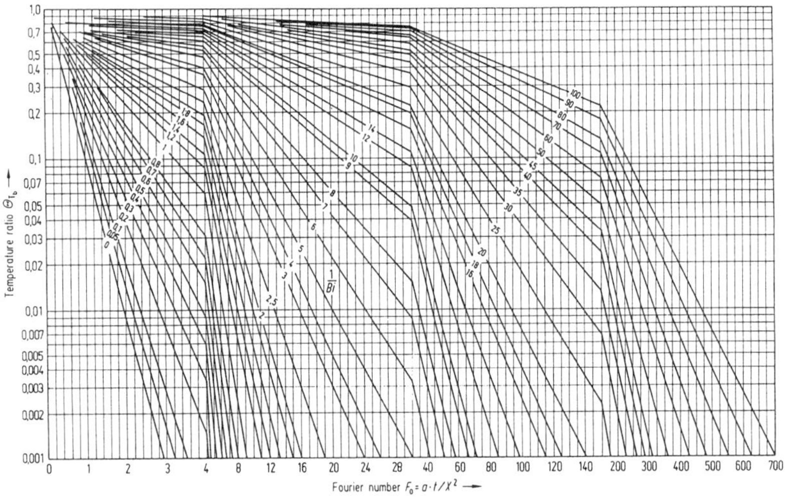

Fig. 7.31: Midplane temperature for an infinite plate [29]

The Fourier number Fo for $\theta_{T_b} = 0.706$ and $1/Bi = 0.0846$ from Fig. 7.31 is approximately Fo = 0.16.

The cooling time t_k follows from the definition of the Fourier number

$$\frac{a \cdot t_k}{R^2} = \frac{10^{-7} \cdot t_k}{(1.6 \cdot 10^{-3})^2} = \frac{t_k}{2.56 \cdot 10} = 0.16$$

$t_k = 4.1$ s

The length of the water bath is therefore

$V_H \cdot t_k = 0.5 \cdot 4.1 = 2.05$ m

If the wire is cooled by spraying water on to ist as it emerges from the extruder, a heat transfer coefficient of $\alpha a = 3500$ W / ($m^2 \cdot$ k) can be assumed.

The Biot number is then

$Bi = 3500 \cdot 1.6 / (1000 \cdot 0.23) = 24.35$

$1/Bi = 0.041$

The Fourier number from Fig. 7.31 is approximately

$F_o \approx 0.15$

The cooling time t_k follows from

$a \cdot t_k / R^2 = 10^{-7} \cdot t_k / (1.6 \cdot 10^{-3})^2 = 0.15$

$t_k = 3.84$ s

This value is not significantly less than the one in the case of cooling in the water bath, since the resistance to heat flow is mainly due to conduction.

Example:
Cooling of a blown film for the following conditions:

thickness of the film:	70 μ
melt temperature:	220 °C
frost line temperature:	110 °C
air temperature:	20 °C
αa	$= 120\ W / (m^2 \cdot K)$
$\lambda_{plastic}$	$= 0.242\ W / (m \cdot K)$
thermal diffusivity a:	$1.29 \cdot 10^{-3}\ cm^2/s$
haul-off speed:	60 m/min

The cooling time t_k is to be calculated.

Solution:
Temperature ratio Θ_{Tb}

$$\Theta_{Tb} = (110 - 20) / 220 - 20) = 0.45$$

Biot number Bi

$$Bi = 120 \cdot 35 \cdot 10^{-6} / 0.242 = 0.01736$$

$$1 / Bi = 57.6$$

F_o from Fig. 7.31 for $\Theta_{Tb} = 0.45$ and $1 / Bi = 57.6$

$$F_o = 42.5$$

$$\text{Cooling time } t_k = \frac{x^2 \cdot F_o}{a} = \frac{35 \cdot 35 \cdot 42.5 \cdot 10^3}{10^8 \cdot 1.29} = 0.4 \text{ s}$$

At a haul-off speed of 1 m/s the height of the film to the frost line will be approximately 0.4 m.

Example:
Cooling of a sheet for the following conditions:

thickness of the sheet	$s = 0.25$ mm
melt temperature	$T_o = 240$ °C
roll temperature	$T_w = 20$ °C
thermal diffusivity	$a = 1.13 \cdot 10^{-3}\ m^2/s$
	$\lambda_{plastic} = 0.12\ W / (m^2 \cdot K)$
	$\alpha_a = 1000\ W / (m^2 \cdot K)$

haul-off speed = 130 m/min

The length of contact between the sheet and the roll required to attain a sheet temperature of 70 °C is to be calculated.

Solution:

The Biot number $\text{Bi} = \dfrac{\alpha_a\, x}{\lambda}$

$\text{Bi} = 1000 \cdot 0.125 / 0.12 = 1.04$

$1 / \text{Bi} = 0.96$

$\Theta_{Tb} = (70 - 120) / 240 - 20) = 50 / 220 = 0.227$

The Fourier number F_o for $1 / \text{Bi} = 0.96$ and Θ_{Tb} from Fig. 7.31

$F_o \approx 2$

Cooling time $t_k = \dfrac{0.125 \cdot 0.125 \cdot 2 \cdot 10^3}{10^2 \cdot 1.13} = 0.276\ \text{s}$

At a haul-off speed of 2.17 m/s the length of contact is L = 2.17 · 0.276 = 0.6 m. For a roll diameter of D = 250 mm this amounts to 1.31 times the roll circumference.

The above examples give only a rough estimate of the actual values. More realistic results can be obtained by using the software POLYFLOW [20].

7.6.1 Dimensionless Groups

Dimensionless groups can be used to describe complicated processes which are influenced by a large number of variables with the advantage that the whole process can be analyzed on a sound basis by means of a few dimensionless parameters.

The foregoing example shows their application in heat transfer applications. Their use in correlating experimental data and in scaling-up equipment is well known.

Table 7.18 shows some of the dimensionless groups which are often used in plastics engineering.

Table 7.18: Dimensionless Groups

Symbol	Name	Definition
Bi	Biot number	$\alpha_a \cdot l/\lambda_i$
Br	Brinkman number	$\eta w^2/(\lambda \cdot \Delta T)$
Deb	Deborah number	t_D/t_p
Fo	Fourier number	$a \cdot t/l^2$
Gr	Grashof number	$g \cdot \beta \cdot \Delta T \cdot l^2/\nu^2$
Gz	Graetz number	$l^2/(a \cdot t_v)$
Le	Lewis number	a/δ_d
Na	Nahme number	$\beta_T \cdot w^2 \cdot \eta/\lambda$
Nu	Nusselt number	$\alpha l/\lambda$
Pe	Peclet number	wl/a
Pr	Prandtl number	ν/a
Re	Reynolds number	$\rho wl/\eta$
Sh	Sherwood number	$\beta_s \cdot l/\delta_d$
Sc	Schmidt number	ν/δ_d
Sk	Stokes number	$p \cdot l/(\eta \cdot w)$

Nomenclature:

a : thermal diffusity [m²/s]
g : acceleration due to gravity [m/s²]
l : characteristic length
p : pressure [N/m²]
t_D : memory time [s]
t_p : process time [s]
ΔT : temperature difference [K]
w : velocity of flow [m/s]
α_a : outside heat-transfer coefficient [W/(m² · K)]
β : coefficient of volumetric expansion [K⁻¹]
β_T : temperature coefficient in the power law of viscosity [1] [K⁻¹]
β_s : mass transfer coefficient [m/s]
δ_d : diffusion coefficient [m²/s]
η : viscosity [Pa · s]
λ : thermal conductivity (index i refers to the inside value) [W/(m · k)]
ν : kinematic viscosity [m²/s]
t_v : residence time [s]
t : time [s]
ρ : density [kg/m³]

Literature

1. *Rao, N.S.:* Design Formulas for Plastics Engineers. Hanser, Munich 1991
2. *Tadmor, Z., Klein, I.:* Engineering Principles of Plasticating Extrusion, Van Nostrand Reinhold, New York (1970)
3. *Bernhardt, E.C.:* Processing of Thermoplastic Materials, Reinhold, New York (1963)
4. *Rao, N.S.:* Computer Aided Design of Plasticating Screws, Hanser, Munich 1986
5. *Klein, I., Marshall, D.I.:* Computer Programs for Plastic Engineers, Reinhold, New York 1968
6. *Rao, N.S.:* Designing Machines and Dies for Polymer Processing with Computer Programs, Hanser, Munich 1981
7. *Procter, B.:* SPE J. 28 (1972) p. 324
8. *Brochure*: EMS Chemie 1992
9. *Brochure: BASF A.G.* 1992
10. *Chung, C.I., Lohkamp, D.T.:* SPE 33, ANTEC 21 (1975), p. 363
11. *Fritz, H.G.:* Extrusion blow molding in Plastics Extrusion Technology, Ed. F. Hensen, Hanser, Munich, 1988
12. *Predöhl, W., Reitemeyer, P.:* Extrusion of pipes, profiles and cables in Plastics Extrusion Technology, Ed. F. Hensen, Hanser, Munich, 1988
13. *Winkler, G.:* Extrusion of blown films in Plastics Extrusion Technology, Ed. F. Hensen, Hanser, Munich, 1988
14. *Bongaerts, H.:* Flat film extrusion using chill-roll casting in Plastics Extrusion Technology, Ed. F. Hensen, Hanser, Munich, 1988
15. *Rauwendaal, C.J.:* Polymer Engineering and Science, 21 (1981), p. 1092
16. *Darnell, W.H., and Mol, E.A.J.:* Soc. Plastics Eng. J., 12, 20 (1956)
17. *Michaeli, W.:* Extrusion Dies, Hanser, Munich 1992
18. *Schenkel, G.:* Private Communication
19. *Squires, P.H.:* SPE-J. 16 (1960), p. 267
20. *Rao, N.S., O'Brien, K.T. and Harry, D.H.:* Computer Modelling for Extrusion and Other Continuous Polymer Processes. Ed. Keith T. O`Brien, Hanser, Munich 1992
21. *Eckstein, Y., Jackson, R.L.:* Plastics Engineering, 5(1995)
22. *Ogorkiewicz, R.M.:* Thermoplastics - Effects of Processing. Gliffe Books Ltd, London 1969
23. *Titow, W.V.:* PVC Technology, Elsevier Applied Science Publishers, London 1984
24. *Rosato, D.V., Rosato, D.V.:* Plastics Processing Data Handbook, Van Nostrand-Reinhold, New York 1990
25. *Hermann, H.:* Compound Lines in Plastics Extrusion Technology, Ed. F. Hensen, Munich 1988
26. *Hensen, F.:* Plastics Extrusion Technology, Hanser, Munich 1988
27. *Rauwendaal, C.:* Polymer Extrusion, Hanser, Munich 1986
28. *Ogorkiewicz, R.M.:* Thermoplastics and Design, John Wiley, New York 1973
29. *Kreith, F., Black, W.Z.:* Basic Heat Transfer, Harper & Row, New York 1980

8 Blow Molding

8.1 Processes

The different processes of blow molding, namely extrusion blow molding, injection blow molding and stretch blow molding are illustrated in the Fig. 8.1 to 8.4 [1], [2]. In Fig. 8.5 [2], in addition to the aforementioned processes the principle of dip molding is briefly explained.

8.1.1 Resin-dependent Parameters

Melt Temperature and Pressure

Typical values of melt temperature and melt pressure for extrusion blow molding are given in Table 8.1.

Table 8.1: Melt Temperature and Pressure for Extrusion Blow Molding of Some Polymers

Material	Melt temperature °C	Melt pressure bar
PE-LD	140/150	100/150
PE-HD	160/190	100/200
PP	230/235	150/200
PVC-U	180/210	100/200
PVC-P	160/165	75/150

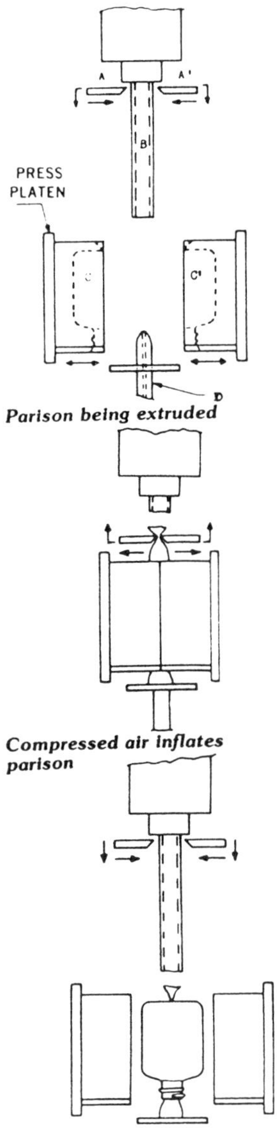

Fig. 8.1: Extrusion blow molding [1]

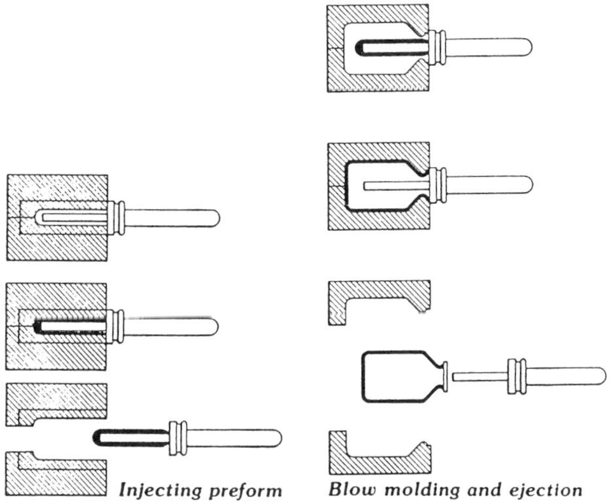

Fig. 8.2: Injection blow molding [1]

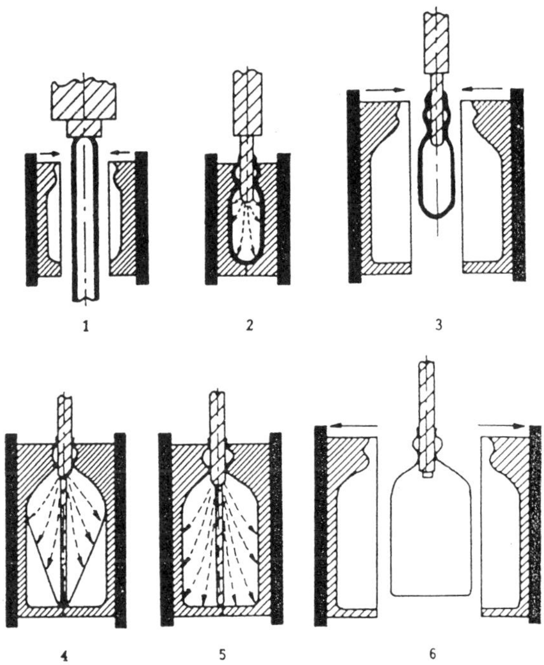

Fig. 8.3: Extrusion stretch blow molding [2]

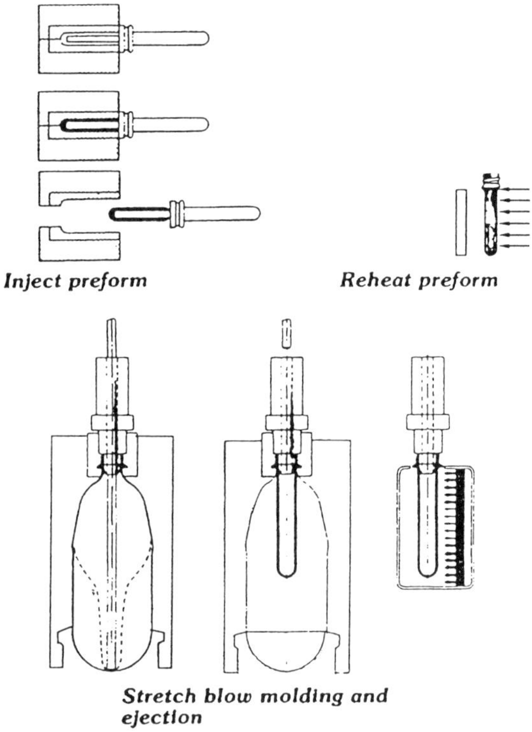

Fig. 8.4: Injection stretch blow molding [2]

Parison Swell

The wall thickness of the molding is related to the swelling ratio of the parison. Referring to Fig. 8.11 [3] the swelling thickness of the parison is given by

$$B_t = h_p/h_d \tag{8.1.1}$$

and the swelling of the parison diameter

$$B_p = D_p/D_d \tag{8.1.2}$$

Using the relationship [4]

$$B_t = B_p^2 \tag{8.1.3}$$

it follows

$$h_p = h_d \cdot B_p^2$$

The swell ratio B_p depends on recoverable strain [4], and can be measured.

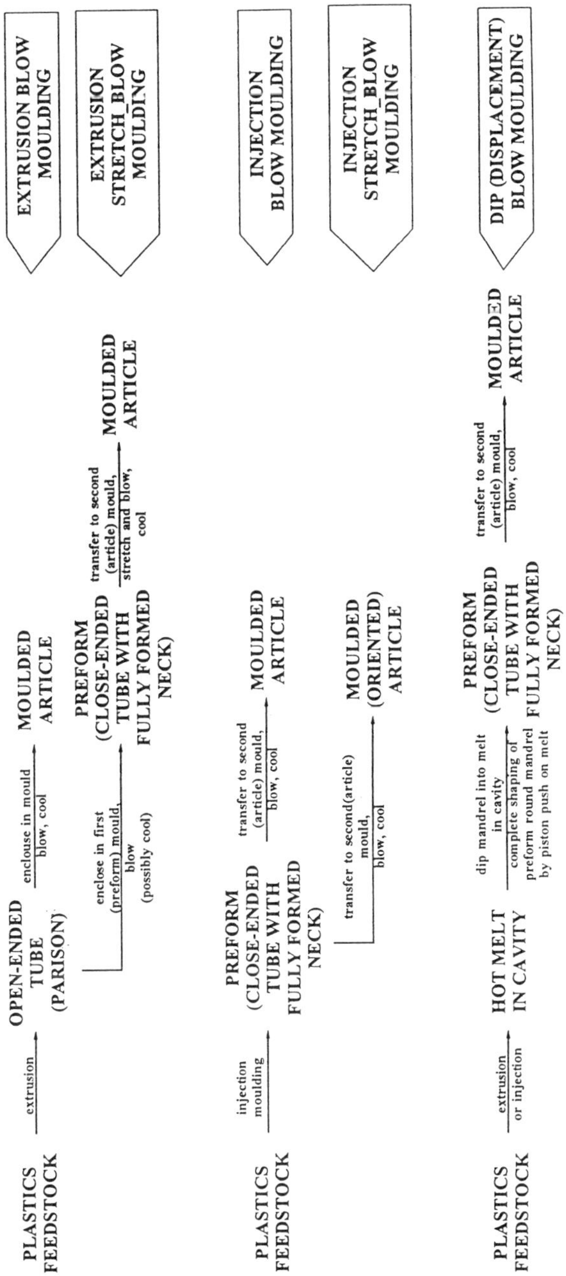

Fig. 8.5: Sequence of operations in different blow molding processes [2]

Average parison swell for some polymers is given in Table 8.2 [1].

Table 8.2: Average Parison Swell for Some Polymers [1]

Polymer	Swell %
PE-HD (Phillips)	15/40
PE-HD (Ziegler)	25/65
PE-LD	30/65
PVC-U	30/35
PS	10/20
PC	5/10

Processing Data for Stretch Blow Molding

Table 8.3: Data on Stretch Blow Molding for Some Polymers [1]

Polymer	Melt temperature °C	Stretch orientation temperature °C	Maximum stretch ratio
PET	250	88/116	16 : 1
PVC	200	99/116	7 : 1
PP	170	121/136	6 : 1
PAN	210	104/127	9 : 1

Volume Shrinkage

Table 8.4: Volume Shrinkage of Stretch Blow Molded Bottles (Seven Days at 27°C) [1]

Type of bottle	Percent
Extrusion blow molded PVC	-
Impact-modified PVC (high orientation)	4.2
Impact-modified PVC (medium orientation)	2.4
Impact modified PVC (low orientation)	1.6
Non-impact modified PVC (high orientation)	1.9
Non-impact modified PVC (medium orientation)	1.2
Non-impact modified PVC (low orientation)	0.9
PET	1.2

Choice of Material

Table 8.5: Data for Choosing Blow Molding Materials [8]

Polymer	Appearance in blow molded form	Cost for equivalent stiffness (PE-LD = 100)
High-impact PVC-U	clear high gloss	104
PE-LD	transculent, high gloss	100
Low-impact PVC-U	very clear, high gloss	87
High-impact PP	rather opaque, low gloss	79
PE-HD	rather opaque, low gloss	76
Low-impact PP	transculent, moderate gloss	71

8.1.2 Machine Related Parameters

Blow Molding Dies

As the parison is always ejected downward and the position of the extruder being generally horizontal, the term cross-head dies may be applied to blow molding dies. These can be of the spider type (Fig. 8.6) [5] or side fed dies as shown schematically in Fig. 7.26 [6]. As mentioned earlier, these dies can be designed by using the equations given in Section 7.3.1.

The interaction of various factors influencing blow molding operations is presented in Figs. 8.7-8.9 [7] and Fig. 8.10 [8]. Data on air blowing pressures and temperatures for cavities in blow molds are given in Tables 8.6 and 8.7 respectively.

Table 8.6: Data on Air Blowing Pressures [1]

Polymer	Pressure bar
Acetal	6.9/10.34
PMMA	3.4/5.2
PC	4.8/10.34
PE-LD	1.38/4.14
PE-HD	4.13/6.9
PP	5.2/6.9
PS	2.76/6.9
PVC-U	5.2/6.9
ABS	3.4/10.34

Table 8.7: Recommended Temperature for Cavities in Blow Molds [1]

Polymer	Temperature °C
PE and PVC	15/30
PC	50/70
PP	30/60
PS	40/65
PMMA	40/60

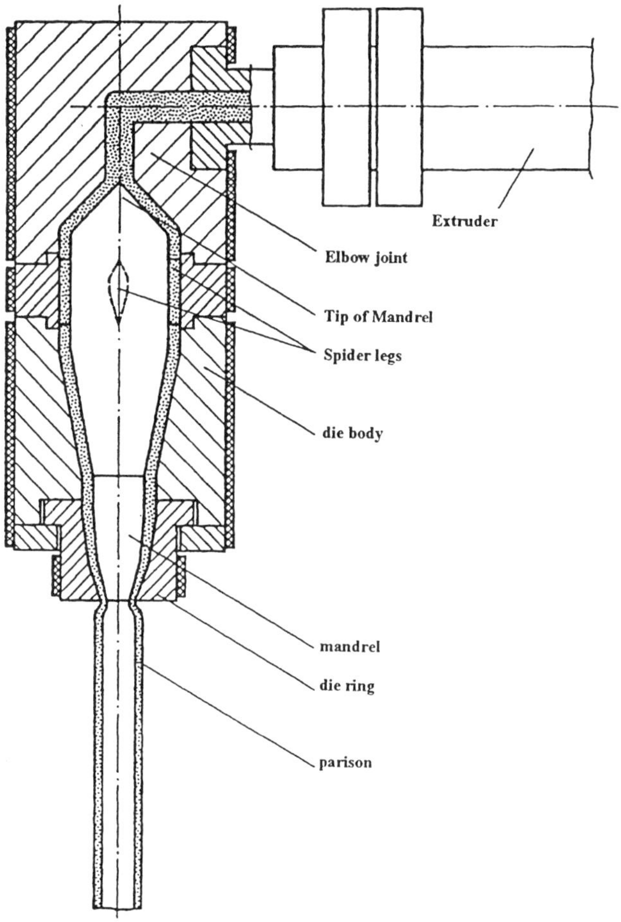

Fig. 8.6: Application of spider die in blow molding [3]

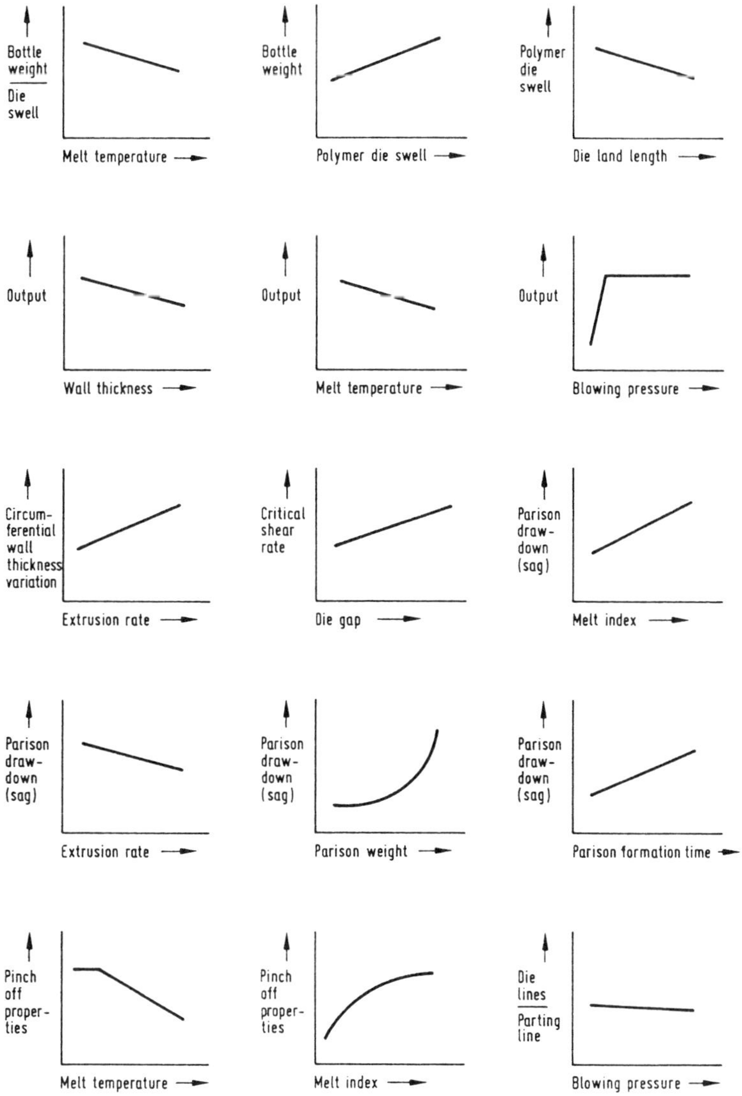

Fig. 8.7: Relationship between machine and material variables in extrusion blow molding process [7]

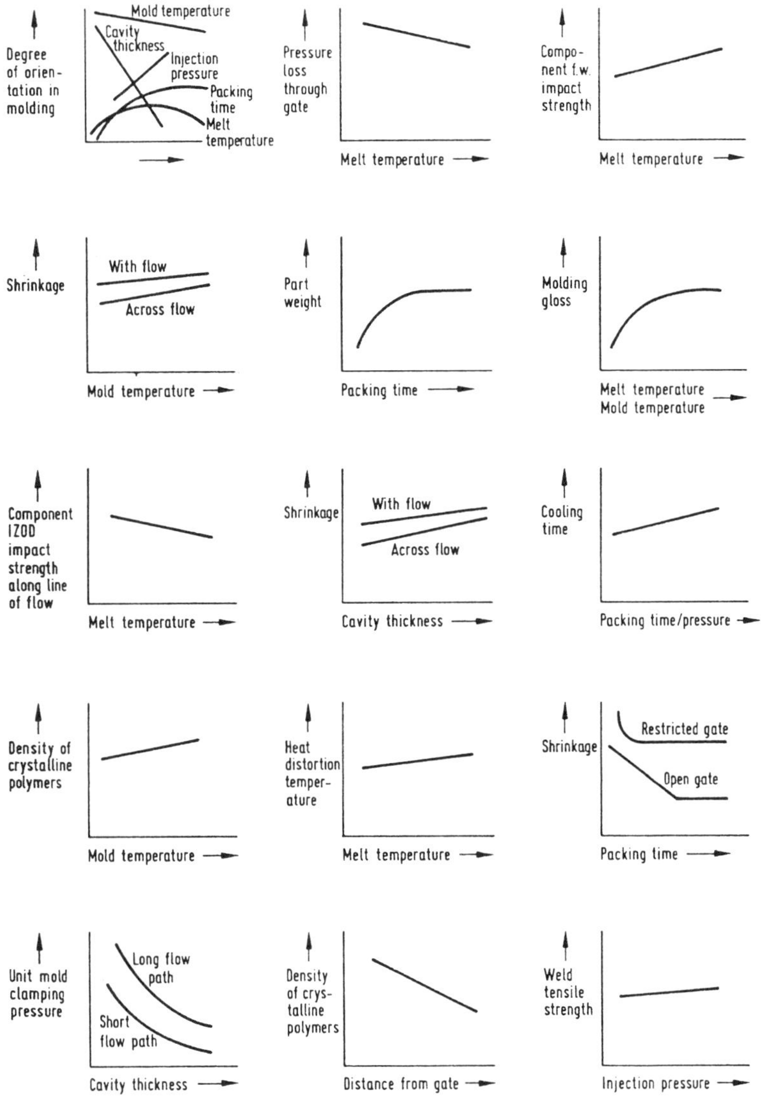

Fig. 8.8: Relationship between machine and material variables in injection blow molding process [7]

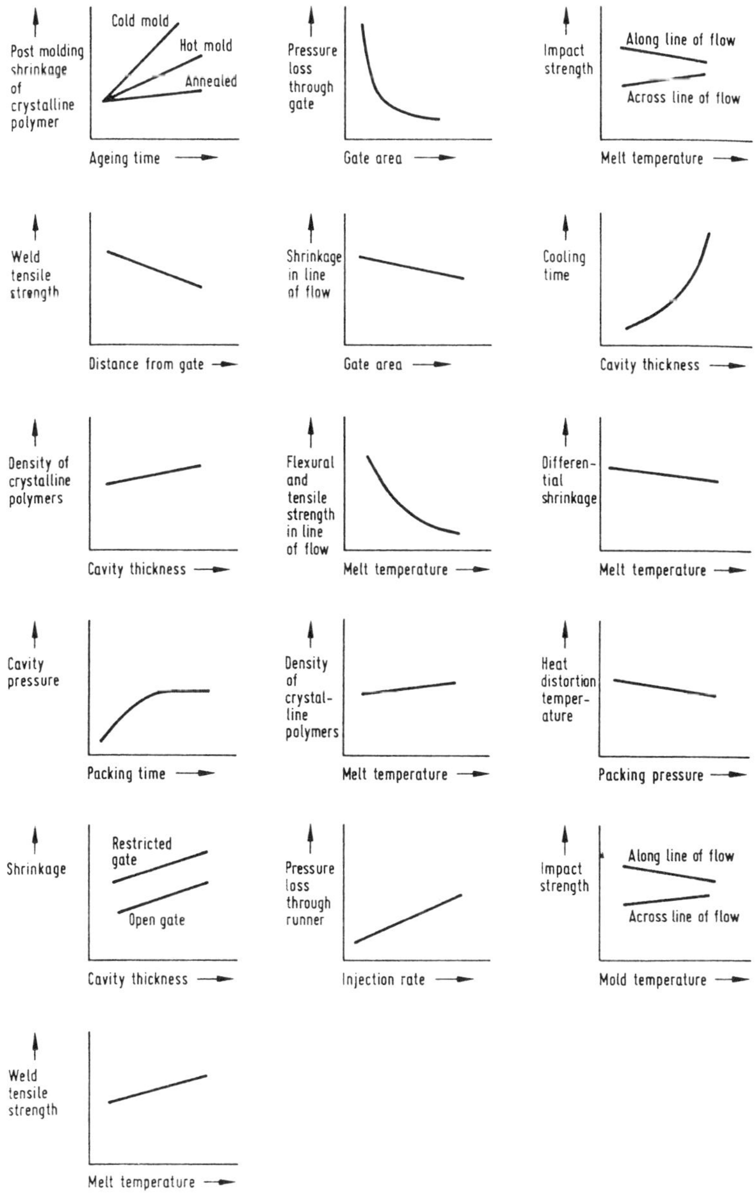

Fig. 8.9: Relationship between machine and material variables in injection blow molding process

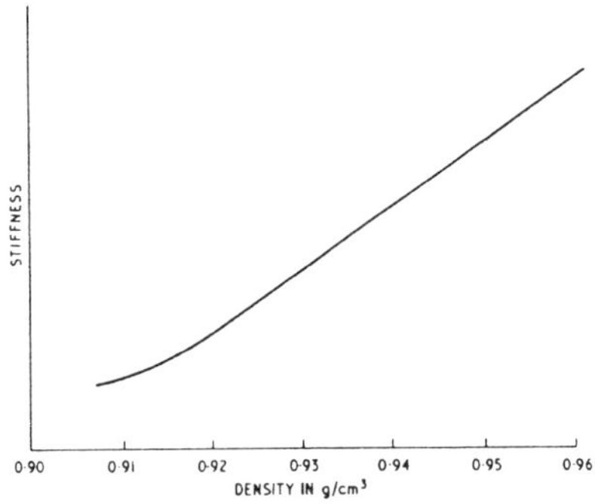

Variation of stiffness with density for polyethylene

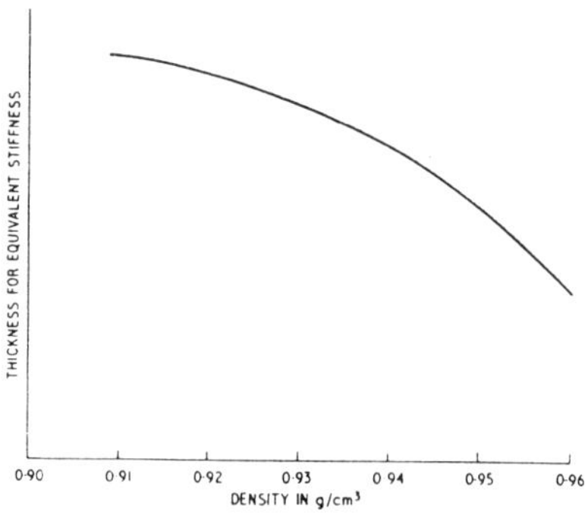

Relationship between thickness for equivalent stiffness and density for polyethylene

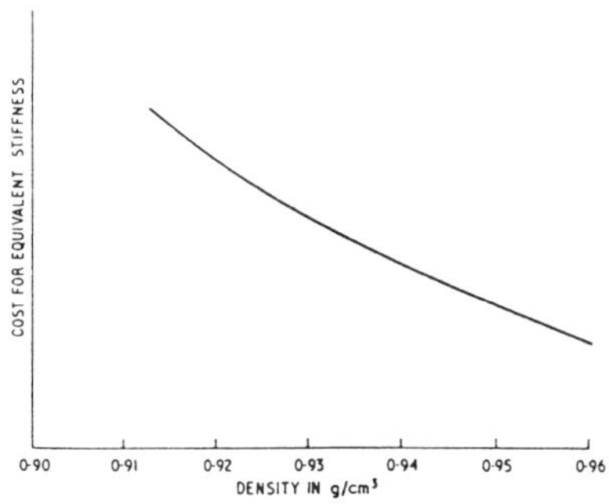

Relationship between cost for equivalent stiffness and density for polyethylene

Fig. 8.10: Relationship between product stiffness, material density and cost in blow molding process [8]

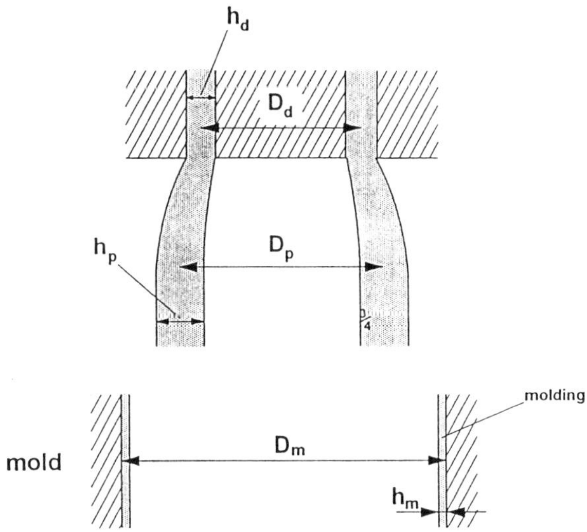

Fig. 8.11: Wall thickness of parison and molding in blow molding [3]

Literature

1. *Rosato, D.V., Rosato, D.V.:* Plastics Processing Data Handbook, Van Nostrand-Reinhold, New York 1990
2. *Titow, W.V.:* PVC Technology, Elsevier Applied Science Publishers, London 1984
3. *Morton-Jones, D.H.:* Polymer Processing, Chapman and Hall, New York 1981
4. *Cogswell, F.N.:* Polymer Melt Rheology, John Wiley, New York 1981
5. *BASF Brochure on Blow Molding* 1970
6. *Michaeli, W.:* Extrusion Dies, Hanser, Munich 1992
7. *Rosato, D.V., Rosato, D.V.:* Injection Molding Handbook, Van Nostrand-Reinhold, New York 1986
8. *Glyde, B.S.:* Blow Moulding in Thermoplastics, Effects of Processing, Ed. R.M. Ogorkiewicz, Gliffe Books Ltd, London 1969

9 Injection Molding

Among the polymer processing operations injection molding (Fig. 9.1) has found the widest application for making articles which can be put to direct use.

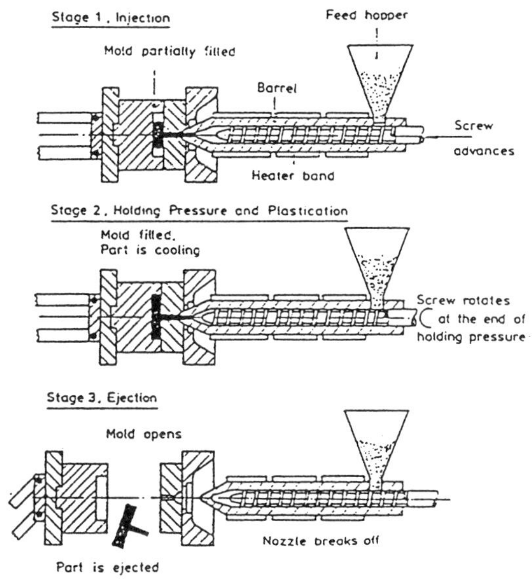

Fig. 9.1: The three stages of injection molding: injection, plastication, ejection [1]

The molding equipment consists of an injection unit, an injection mold and a mold temperature control unit. The quality of the part depends on how well the interaction between these components functions.

At an appropriate point in the typical molding cycle (Fig. 9.1) the screw remains stationary in the forward position (Fig. 9.1A) as indicated by phase 5 in Fig. 9.1B [22]. At the end of dwell or holding pressure the screw begins to rotate conveying plastic material and developing a pressure ahead of the screw. This pressure pushes the screw to move back a predetermined distance which is dependent on the desired volume of the molded part. The screw is then idle in the back position while the previously molded plastic continues cooling in the mold, and while the mold is open and part ejected. After the mold closes, the screw is forced forward by the hydraulic pressure, causing the newly recharged shot in front of the screw to flow into the empty mold. A valve, such as a cheek ring (Fig. 9.7A), prevents back flow during injection. The screw then maintains pressure on the molded plastic for a specific time called dwell or hold time, thus completing the cycle [7].

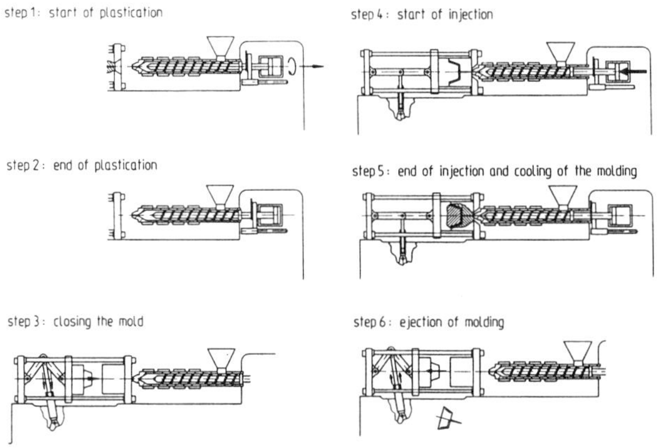

Fig. 9.1A: Injection molding process [22]

The important processing variables in an injection molding process and their effect on the plastic material and part are shown in Fig. 9.1C, in which the cavity pressure is illustrated as a function of time during the molding of a part [1].

As in extrusion, the processing parameters can be classified into resin-dependent and machine-related parameters.

9.1 Resin-Dependent Parameters

9.1.1 Injection Pressure

Injection pressure is the pressure exerted on the melt in front of the screw tip during the injection stage, with the screw acting like a plunger [1]. It affects both the speed of the advancing screw and the process of filling the mold cavity with the polymer melt, and corresponds to the flow resistance of the melt in the nozzle, sprue, runner and cavity. The injection pressures needed for some resins are given in Table 9.1 [1]. The maximum injection pressure is approximately 1.2 times the pressure listed in Table 9.1.

Table 9.1: Injection Pressure Required for Various Plastics [1]

Material	Necessary injection pressure [MPa]		
	Low viscosity heavy sections	Medium viscosity standard sections	High viscosity thin sections, small gates
ABS	80/110	100/130	130/150
POM	85/100	100/120	120/150
PE	70/100	100/120	120/150
PA	90/110	110/140	> 140
PC	100/120	120/150	< 150
PMMA	100/120	120/150	< 150
PS	80/100	100/120	120/150
Rigid PVC	100/120	120/150	> 150
Thermosets	100/140	140/175	175/230
Elastomers	80/100	100/120	120/150

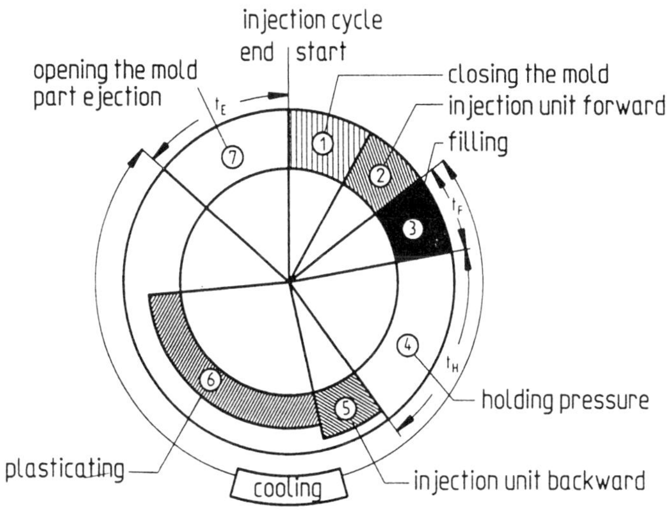

Fig. 9.1B: Molding cycle [22]

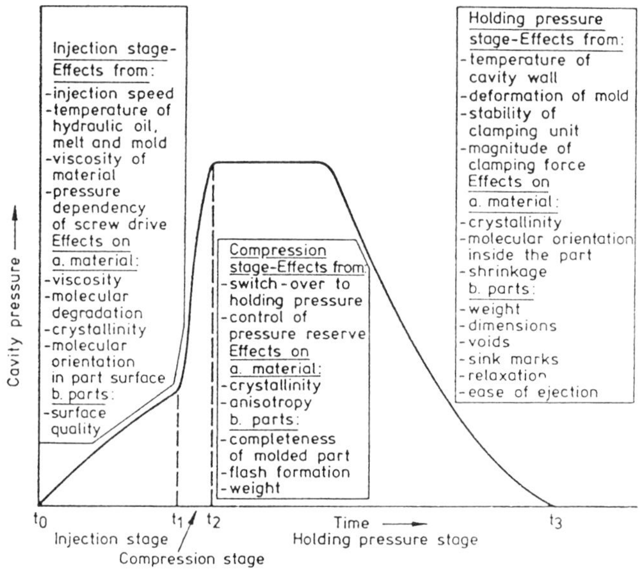

Fig. 9.1C: Cavity pressure profile over time [1]

9.1.2 Mold Shrinkage and Processing Temperature

Due to their high coefficients of thermal expansion components molded from plastics experience significant shrinkage effects during the cooling phase. Semi-crystalline polymers tend to shrink more, owing to the greater specific volume difference between melt and solid. Shrinkage S can be defined by [2]

$$S = \frac{L^* - L}{L^*} \tag{9.1.1}$$

where L^* is any linear dimension of the mold and L, the corresponding dimension of the part when it is at some standard temperature and pressure. As a consequence of mold shrinkage it is necessary to make core and cavity slightly larger in dimension than the size of the finished part.

Processing temperature, mold temperature and shrinkage are given in Table 9.2 for a number of materials [1].

The parameters influencing shrinkage behavior are shown in Fig. 9.2 [17].

Table 9.2: Processing Temperatures, Mold Temperatures, and Shrinkage for the Most Common Plastics Processed by Injection Molding [1]

Material	Glass fiber content %	Processing temperature °C	Mold temperature °C	Shrinkage %
PS		180/280	10/10	0.3/0.6
HI-PS		170/260	5/75	0.5/0.6
SAN		180/270	50/80	0.5/0.7
ABS		210/275	50/90	0.4/0.7
ASA		230/260	40/90	0.4/0.6
PE-LD		160/260	50/70	1.5/5.0
PE-HD		260/300	30/70	1.5/3.0
PP		250/270	50/75	1.0/2.5
PPGR	30	260/280	50/80	0.5/1.2
IB		150/200		
PMP		280/310	70	1.5/3.0
PVC-soft		170/200	15/50	> 0.5
PVC-rigid		180/210	30/50	- 0.5
PVDF		250/270	90/100	3/6
PTFE		320/360	200/230	3.5/6.0
FEP				
PMMA		210/240	50/70	0.1/0.8
POM		200/210	> 90	1.9/2.3
PPO		250/300	80/100	0.5/0.7
PPO-GR	30	280/300	80/100	< 0.7
CA		180/230	50/80	0.5
CAB		180/230	50/80	0.5
CP		180/230	50/80	0.5
PC		280/320	80/100	0.8
PC-GR	10/30	300/330	100/120	0.15/0.55
PET		260/290	140	1.2/2.0
PET-GR	20/30	260/290	140	1.2/2.0
PBT		240/260	60/80	1.5/2.5
PBT-GR	30/50	250/270	60/80	0.3/1.2
PA 6		240/260	70/120	0.5/2.2
PA 6-GR	30/50	270/290	70/120	0.3/1
PA 66		260/290	70/120	0.5/2.5
PA 66-GR	30/35	280/310	70/120	0.5/1.5
PA 11		210/250	40/80	0.5/1.5
PA 12		210/250	40/80	0.5/1.5
PSO		310/390	100/160	0.7
PPS	40	370	> 150	0.2
PUR		195/230	20/40	0.9
PF		60/80	170/190	1.2
MF		70/80	150/165	1.2/2
MPF		60/80	160/180	0.8/1.8
UP		40/60	150/170	0.5/0.8
EP	30/80	ca. 70	150/170	0.2

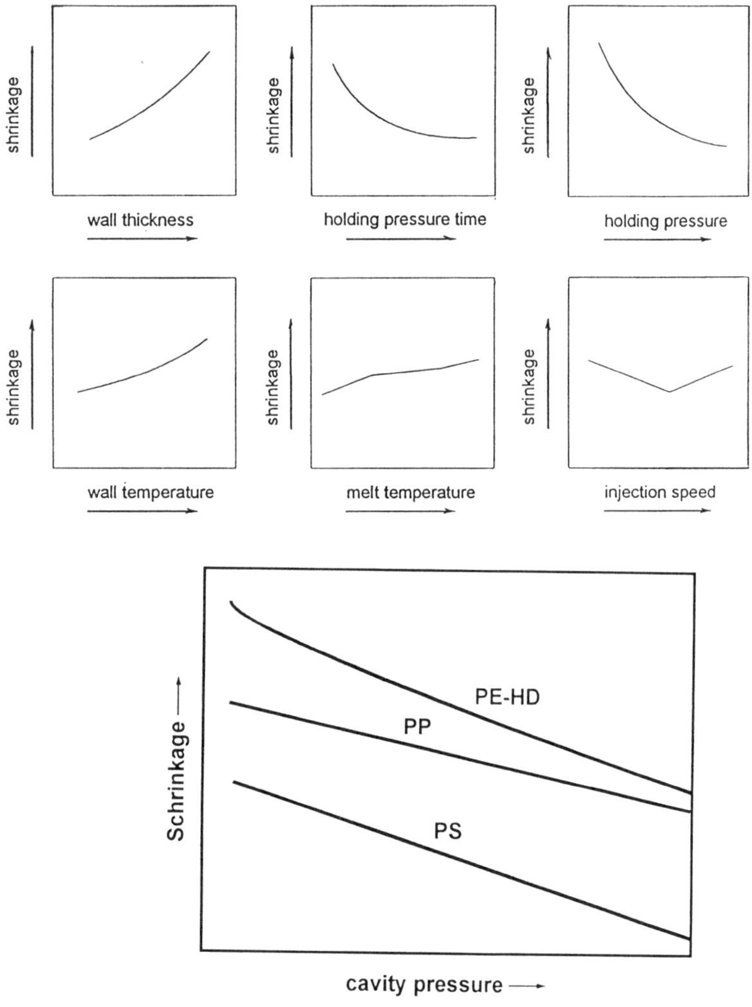

Fig. 9.2: Qualitative relations between individual process parameters and shrinkage behavior [17]

9.1.3 Drying Temperatures and Times

As already mentioned in Section 6.1.4 plastics absorb water at a rate depending on the relative humidity of the air in which they are stored (Table 9.3) [3].

Table 9.3: Equilibrium Moisture Content of Various Plastics Stored in Air at 23°C and Relative Humidity 50% [3]

	CA	CAB	ABS	PA6	PA66	PBT	PC	PMMA	PPO	PET
Equilibrium moisture (%)	2.2	1.3	1.5	3	2.8	0.2	0.19	0.8	0.1	0.15
Water content acceptable for injection molding [%]	0.2	0.2	0.2	0.15	0.03	0.02	0.08	0.05	0.02	0.02

Moisture and monomers have an adverse effect on processing and on the part. To remove moisture, materials are dried in their solid or melt state. Table 9.4 shows the drying temperatures and times for different resins and dryers [1].

Table 9.4: Drying Temperatures and Times of Various Dryers [1]

	Drying temperature °C		Drying time h	
Material	fresh air/mixed air dryers	dehumidifying dryers	fresh air/mixed air dryers	dehumidifying dryers
ABS	80	80	2/3	1/2
CA	70/80	75	1/1.5	1
CAB	70/80	75	1/1.5	1
PA 6	not recomm.	75/80	-	2
PA 66,6.10	"	75/80	-	2
PBT, PET	120	120	3/4	2/3
PC	120	120	2/4	2
PMMA	80	80	1/2	1/1.5
PPO	120	120	1/2	1/1.5
SAN	80	80	1/2	1/1.5

9.1.4 Flow Characteristics of Injection-Molding Resins

The flow behavior of injection-molding materials can be determined on the basis of melt flow in a spiral channel. In practice a spiral-shaped mold of rectangular cross-section with the height and width in the order of a few millimeters is often used to classify the resins according to their flowability (Fig. 9.3). The length L of the solidified plastic in the spiral is taken as a measure of the viscosity of the polymer concerned (Fig. 9.4).

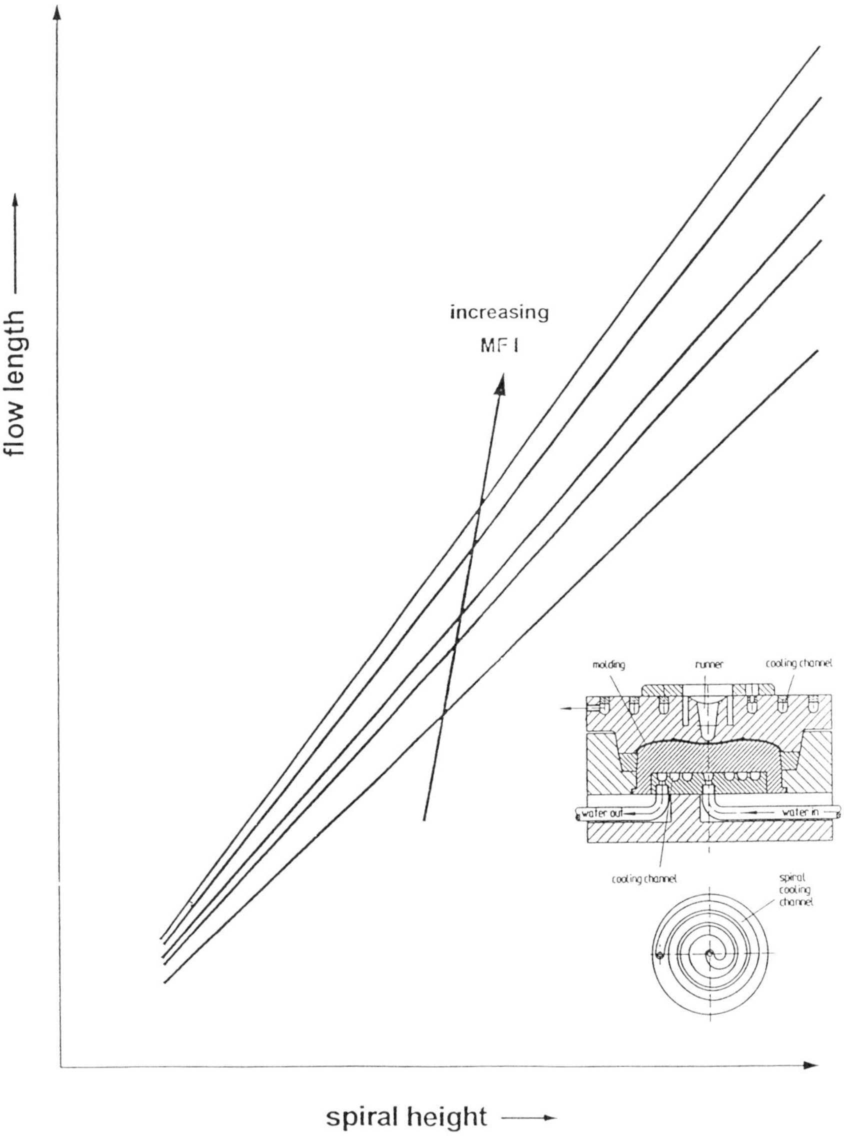

Fig. 9.3: Flow length in a spiral test as a function of the resin MFI

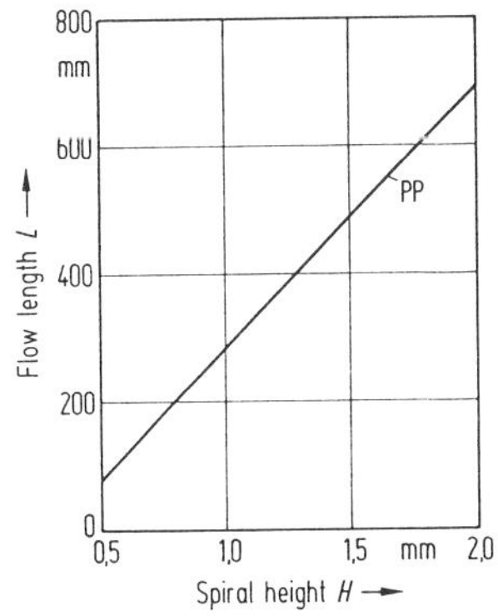

Fig. 9.4: Flow Length L as function of the spiral height H [4]

Fig. 9.4 shows the experimentally determined flow length L as a function of the height H of the spiral for polypropylene. A quantitative relation between L and the parameters influencing L such as type of resin, melt temperature, mold temperature and injection pressure can be developed [4] by using the dimensionless numbers as defined by Thorne [5]. As shown in [4] the Graetz number correlates well with the product Re · Pr · Br.

$$Gz = f(Re \cdot Pr \cdot Br), \qquad (9.1.2)$$

from which an explicit relationship for the flow length L can be computed.

Example [4]:
This example illustrates the calculation of the dimensionless numbers Gz, Re, Pr and Br for the given data and for resin-dependent constants given in [4]:

Width of the spiral	W	=	10 mm
Height of the spiral	H	=	2 mm
Flow length	L	=	420 mm
Melt density	ρ	=	1.06 g/cm^3
Specific heat	c_p	=	2 kJ/(kg · K)
Thermal conductivity	λ	=	1.5 W/(m · K)
Melt temperature	T_M	=	270 °C
Mold temperature	T_W	=	70 °C
Mass flow rate	G	=	211.5 kg/h

Solution:
The conversion factors for the units used in the following calculation of the dimensionless numbers are

$$F_1 = 0.001 \;;\; F_2 = 1000 \;;\; F_3 = 3600$$

The Graetz number Gz is calculated from

$$Gz = \frac{F_2 \cdot G \cdot c_p}{F_1 \cdot F_3 \cdot \lambda \cdot L} \tag{9.1.3}$$

with G in kg/h and L in mm.
From the above data Gz = 186.51.
The Reynolds number is obtained from

$$Re = \frac{F_2 \cdot V_e^{(2-n_R)} \cdot \rho \cdot H^{*\ n_R}}{k^*} \tag{9.1.4}$$

where V_e = velocity of the melt front in m/s
H^* = half-height of the spiral in mm
n_R = reciprocal of the power law exponent
k^* = $\eta\ \dot{\gamma}^{(1-n_R)}$

The viscosity η can be calculated from the melt temperature and shear rate [4].
For the given values Re = 0.03791 [4].
With H^* in m and V_e in m/s we get for the Prandtl number Pr

$$Pr = \frac{F_2 \cdot k^* \cdot c_p \cdot H^{*(1-n_R)}}{\lambda \cdot V_e^{(1-n_R)}} \tag{9.1.5}$$

Pr = 103 302.87

The Brinkman number Br [5]

$$Br = \frac{k^* \cdot V_e^{(1+n_R)} \cdot H^{*(1-n_R)}}{\lambda \cdot (T_M - T_W)} \tag{9.1.6}$$

V_e follows from $V_e = \dfrac{\dot{Q}}{W \cdot H}$ with $\dot{Q} = \dfrac{G}{\rho}$.

Br = 1.9833

Finally the product Re · Pr · Br is 7768.06.

From the plot of Gz vs Re·Pr·Br (Fig. 9.5) [4], which is obtained by measuring the flow length at different values of spiral height, melt temperature and flow rate the Graetz number corresponding to the calculated product of Re·Pr·Br is read and finally the flow length L calculated from this number.

Symbols and units used in the formulas above:

Br	Brinkman number
c_p	Specific heat kJ/kg·K
$\dot{\gamma}$	Shear rate s^{-1}
G	Mass flow rate kg/h
Gz	Graetz number
H	Height of the spiral mm
H^*	Half-height of the spiral mm

L	Length of the spiral mm
n_r	Reciprocal of the power law exponent
Pr	Prandtl number
$\dot{Q}0$	Volume flow rate m^3/s
Re	Reynolds number
T_M	Melt temperature °C
T_W	Mold temperature °C
Ve	Velocity of the melt front m/s
W	Width of the spiral mm
λ	Thermal conductivity W/m·K
ρ	Melt density g/cm^3
η	Melt viscosity Pa·s

In the case of thermosetting resins the flow can be defined as a measure of melt viscosity, gelation rate and subsequent polymerization or cure [18].

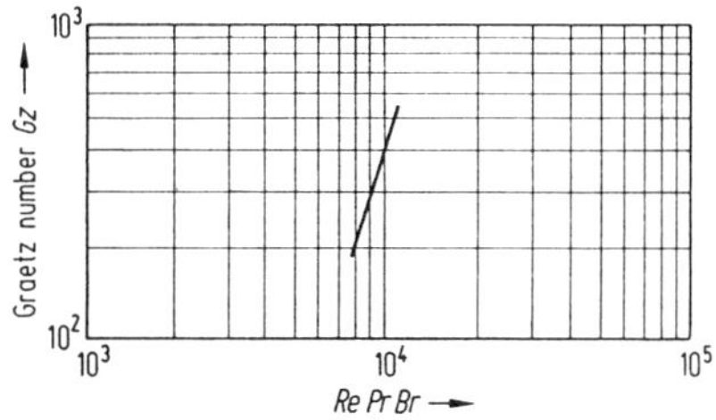

Fig. 9.5: Dimensionless groups for determining the flow length L in s spiral test

9.2 Machine Related Parameters

9.2.1 Injection Unit

The average travel velocities of injection units (Fig. 9.6) [1] are given in Table 9.5 [1]. Table 9.6 [1] shows the common forces with which the nozzle is in contact with the sprue bushing. This contact pressure prevents the melt from leaking into the open at the interface between nozzle and sprue bushing [1].

Table 9.5: Travel Velocity of the Injection Unit [1]

Clamping force kN	Maximum velocity mm/s	Minimum velocity mm/s
< 500	300/400	20/40
501/2000	250/300	30/50
2001/10 000	200/250	40/60
> 10 000	200	50/100

Table 9.6: Contact Force Between Nozzle and Sprue Bushing [1]

Clamping force kN	Contact force kN
500	50/80
1000	60/90
5000	170/220
10 000	220/280
20 000	250/350

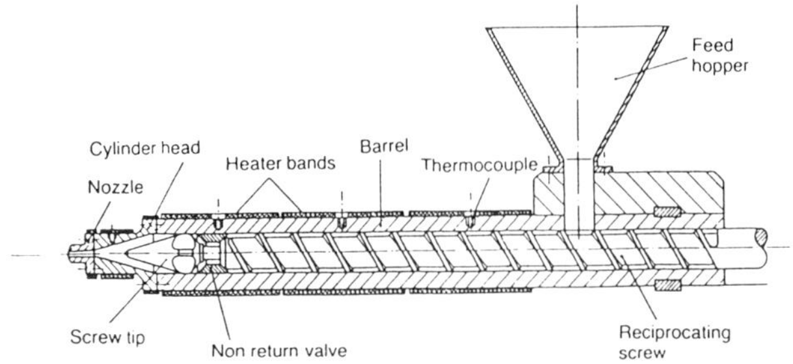

Fig. 9.6: Components of a reciprocating screw injection [1]

9.2.2 Injection Molding Screw

The plastication of solids in the reciprocating screw of an injection molding machine ia a batch process consisting of two phases. During the stationary phase of the screw melting takes place mainly by conduction heating from the barrel. The melting during screw rotation time of the molding cycle is similar to that in an extrusion screw but time-dependent. At long times of screw rotation it approaches the steady state condition of extrusion melting [6], [7].

Design of injection molding screws can be performed by using the software of Rao [8]. Essential dimensions are the lengths and depths of the feed zone and the metering zone (Fig. 9.7). Table 9.7 shows thcsc data for several screw diameters [1].

Table 9.7: Significant Screw Dimensions for Processing Thermoplastics [1]

Diameter	Flight depth (feed) h_F mm	Flight depth (metering) h_M mm	Flight depth ratio	Radial flight clearance mm
30	4.3	2.1	2 : 1	0.15
40	5.4	2.6	2.1 : 1	0.15
60	7.5	3.4	2.2 : 1	0.15
80	9.1	3.8	2.4 : 1	0.20
100	10.7	4.3	2.5 : 1	0.20
120	12	4.8	2.5 : 1	0.25
> 120	max 14	max 5.6	max 3 : 1	0.25

Screw geometries for other polymers are given by Johannaber [1].

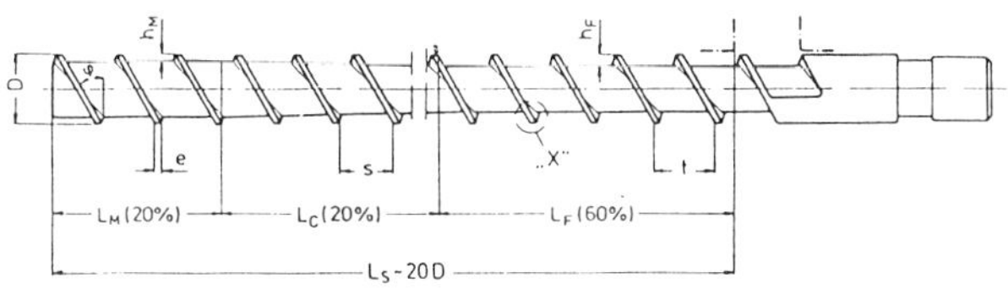

Fig. 9.7: Dimensions of an injection molding screw [1]

Non-return Valves

A non-return valve is a component at the tip of the screw that prevents back flow of the plasticated material during the injection and holding pressure stages. A comparison between the main categories of the non-return valves used (Fig. 9.7A) is given in Table 9.7A [21].

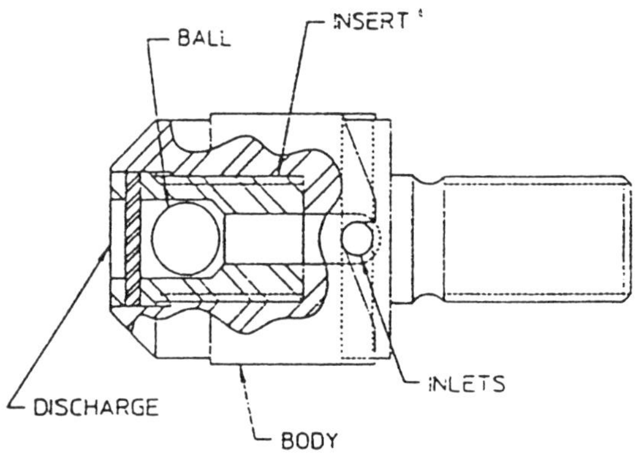

Front Discharge Ball Check

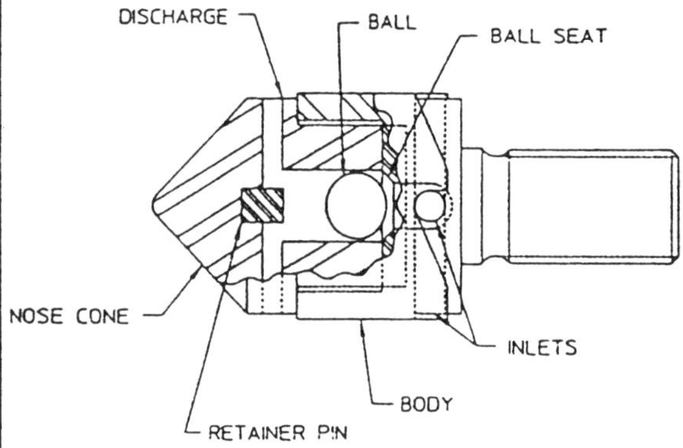

Side Discharge Ball Check

Fig. 9.7A: Non-return values [21], [22]

Table 9.7A: Comparison Between Sliding Ring Valves and Ball Check Valves [21]

Sliding ring valve – Advantages	Ball check valve – Advantages
1. Greater streamling for lower material degradation 2. suitable for heat sensitive materials 3. less barrel wear 4. less pressure drop across valve 5. well suited for vented operation 6. easy to clean.	1. More positive shut off 2. Better shot control
Disadvantages	**Disadvantages**
1. Lower positive shut-off in large machines 2. lower shot control.	1. Less streamlined, hence more degradation of materials 2. more barrel wear and galling 3. greater pressure drop, hence more heat 4. poor for vented operation 5. harder to clean.

Nozzle

The plasticating barrel ends at the mold in a nozzle, which is forced against the sprue bushing of the mold prior to injection and produces a force locking connection there [1]. Examples of commonly used types of nozzles are given in Fig. 9.7B [1], [22].

9.2.3 Injection Mold

The quantitative description of the important mold filling stage has been made possible by the well-known computer programs like MOLDFLOW [9], CADMOULD [10], C-MOLD [19] and others.

The purpose of this section is to touch upon the basic principles of designing injection molding dies using rheological and thermal properties of polymers.

9.2.3.1 Runner Systems

The pressure drop along the gate or runner of an injection mold can be calculated from the same relationships used for dimensioning extrusion dies (Section 7.3).

Example:

A part is to be made from polycarbonate using a runner of length 100 mm and diameter 3 mm and with the melt at a temperature 300 °C. The volume flow rate of the plastic melt through the runner is 2.65 cm^3/s. The pressure drop in the runner is to be calculated.

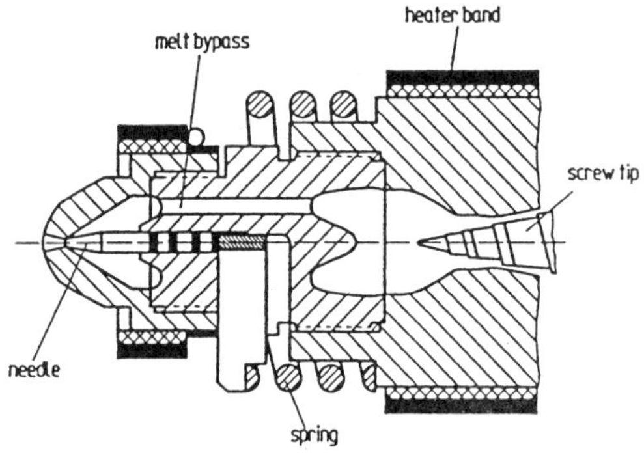

Shutoff nozzle with internal needle (with self-closing spring action)

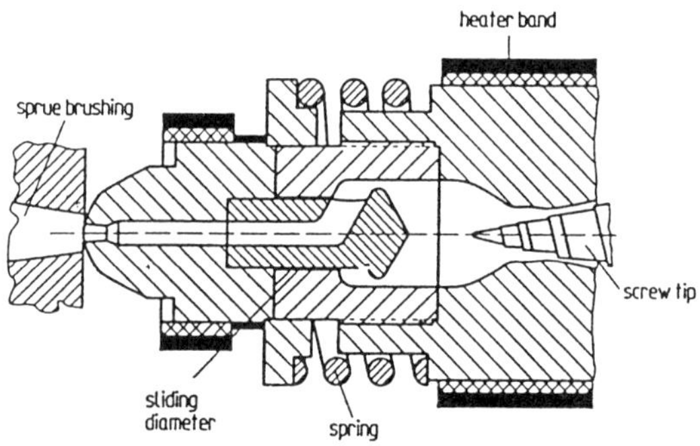

Floating or sliding shutoff nozzle (with self-closing spring action)

Fig. 9.7B: Examples of nozzles [22]

Solution:

shear rate $\dot{\gamma} \;=\; \dfrac{4\dot{Q}}{\pi R^3} \;=\; \dfrac{4 \cdot 2.65 \cdot 10^3}{\pi \cdot 1.5^3} \;=\; 1000 \;\; s^{-1}$

At 300 °C, a shear rate of 1000 s^{-1} corresponds to a viscosity $\eta = 420$ Pa · s according to the resin manufacturer's data.

The shear stress

$$\tau = \eta \cdot \dot{\gamma} = 420 \cdot 1000 = 0.42 \text{ MPa}$$

The pressure drop across the runner can be calculated by using Eq. (3.2.3)

$$\Delta p = \frac{2 \cdot \tau \cdot L}{R} = \frac{2 \cdot 0.42 \cdot 100}{1.5} = 56 \text{ MPa}$$

It is accepted practice that the pressure drop across the runner should be less than 70 MPa [23]. Otherwise the size of the runner should be increased. The pressure drop in runners of non-circular cross-sections can be calculated by using the Eq. (7.2.35) given in Section 7.3.

Different cross-sections for runners, their advantages, disadvantages and practical design relationships are summarized in Fig. 9.8 [20].

Empirical guidelines for dimensioning cross sections of runners are presented in Fig. 9.9 [20].

Circular cross-section	Cross sections for runners	
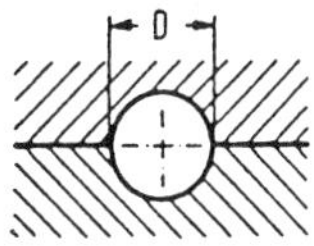 $D = s_{max} + 1{,}5\,mm$	*Advantages:*	smallest surface relative to cross-section, slowest cooling rate, low heat and frictional losses, center of channel freezes last therefore effective holding pressure
	Disadvantages:	Machining into both mold halves is difficult and expensive
Parabolic cross-section 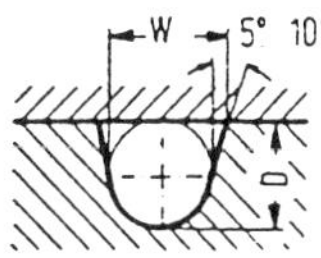 $W = 1{,}25 \cdot D$ $D = s_{max} + 1{,}5\,mm$	*Advantages:*	best approximation of circular cross section, simpler machining in one mold half only (usually movable side for reasons of ejection
	Disadvantages:	more heat losses and scrap compared with circular cross section
Trapezoidal cross-section 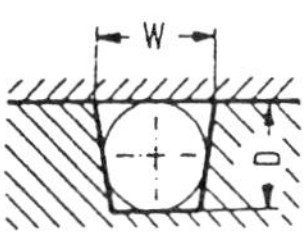 $W = 1{,}25 \cdot D$	Alternative to parabolic cross section	
	Disadvantages:	more heat losses and scrap than parabolic cross section
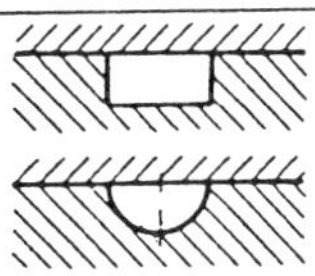	Unfavorable cross sections have to be avoided	

Fig. 9.8: Advantages and disadvantages of some runner cross-sections [20]

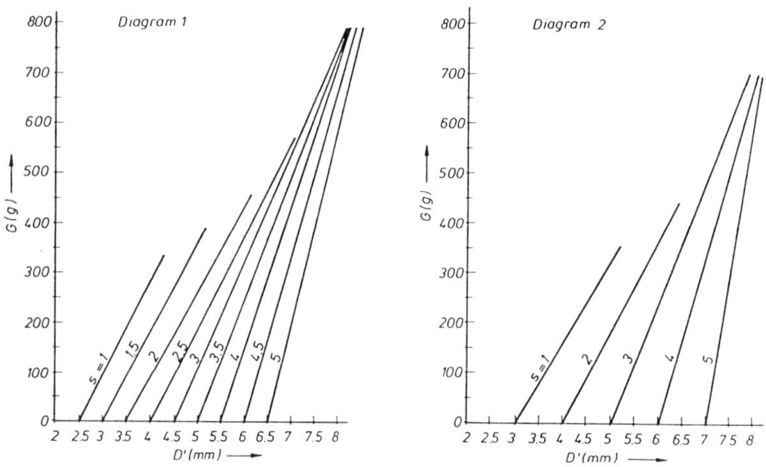

Diagram 1
applicable for PS, ABS, SAN, CAB

Diagram 2
applicable for PE, PP, PA, PC, POM

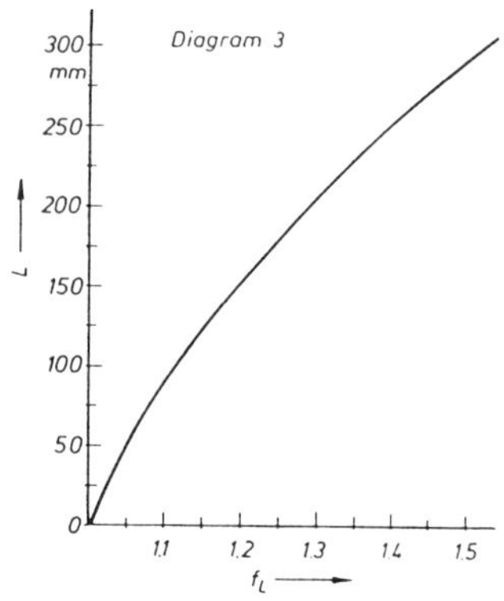

Symbols

s: wall thickness of part (mm)
D′: diameter of runner (mm)
G: weight of part (g)
L: length of runner to one cavity (mm)
f_L: correction factor

Procedure (Diagram 3)

1. determine G and s
2. take D′ from diagram for material considered
3. determine L
4. take f_L from diagram 3
5. correct diameter or runner: $D = D' \cdot f_L$

Fig. 9.9: Nomogram for dimensioning runner cross-sections [20]

9.2.3.2 Design of Gates

Suggested dimensions of some commonly used gates are shown in Fig. 9.10 [20] and those of sprues in Fig. 9.11 [20].

By means of the software mentioned in Section 9.2.3 the entire melt entry system of the mold as well as the mold can be designed with good success.

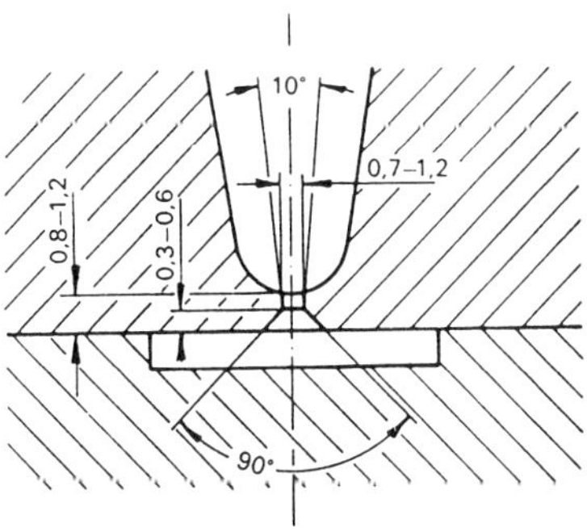

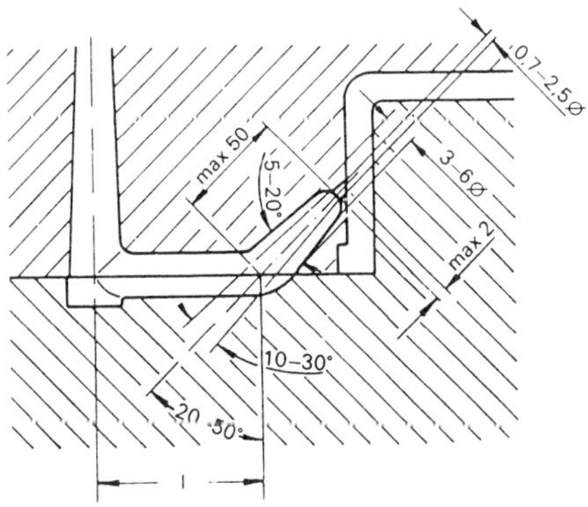

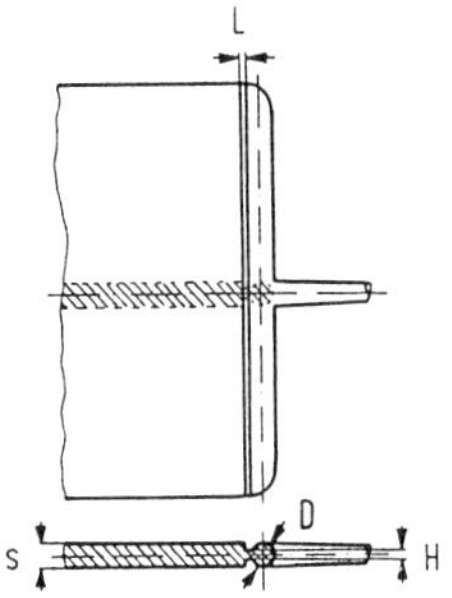

Fig. 175 Edge gate with circular distributor channel [1, 5]
D = s to $\frac{4}{3}$ s + k
k = 2 mm for short flow lengths and thick sections
k = 4 for long flow lengths and thin sections
L = (0.5 to 2.0) mm
H = (0.2 to 0.7) s

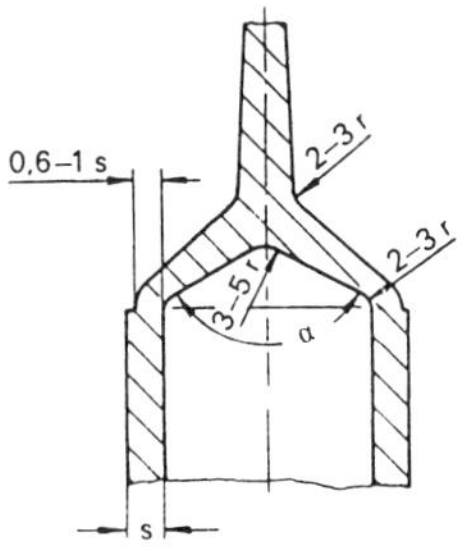

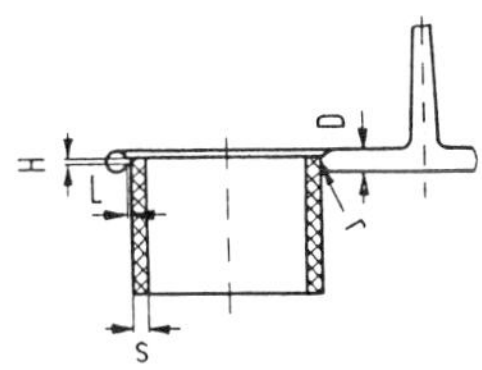

Fig. 185 Ring gate with circular cross section [5, 6]
D = s + 1.5 mm to $\frac{4}{3}$ s + k
L = 0.5 to 1.5 mm
H = $\frac{2}{3}$ s to 1 to 2 mm
r = 0.2 s
k = 2 for short flow lengths and thick sections
k = 4 for long flow lengths and thin sections

Fig. 9.10: Guidelines for dimensioning gates [20] a: pinpoint, b: tunnel, c: edge gate with circular distributor channel, d: disk gate, e: ring gate with circular cross-section

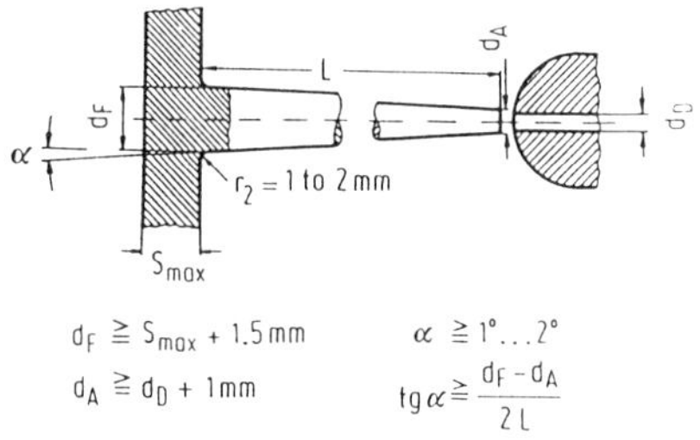

Fig. 9.11: Guidelines for dimensioning a sprue [20]

9.2.3.3 Injection Pressure and Clamp Force

To determine the size of an injection molding machine suited to produce a given part, knowledge of the clamp force required by the mold is important in order to prevent the mold from flashing.

As already mentioned, the mold filling process was treated extensively in the programs [9], and [10] and later on by Bangert [11]. Analytical expressions for the injection pressure and clamp force have been given by Stevenson [12] on the basis of a model for mold filling. The main parameters characterizing this model are fill time τ and Brinkman number Br, which can be used as machine related parameters for describing injection molding in general. According to the solution procedure of Stevenson [12] injection pressure and clamp force are calculated assuming isothermal flow of melt in the cavity and then modified to obtain actual values. The empirical relationships used for this modification are functions of fill time and Brinkman number.

Following *example* shows the calculation of fill time and Brinkman number for a disc-shaped cavity [4], [12]:

The material is ABS with $n_R = 0.2655$, which is the reciprocal of the power law exponent n.

The viscosity constant k_τ [4] is	k_τ	=	$3.05 \cdot 10^4$
Constant injection rate	$\dot{Q}$	=	160 cm^3/s
Part Volume	V	=	160 cm^3
Half-thickness of the disc	b	=	2.1 mm
Radius of the disc	r_2	=	120 mm
Number of gates	N	=	1
Inlet melt temperature	T_m	=	245 °C
Mold temperature	T_w	=	50 °C
Thermal conductivity of the melt	λ	=	0.174 W/(m · K)
Thermal diffusivity of the polymer	a	=	$7.72 \cdot 10^{-4}$ cm^2/s
Melt flow angle [12]	Θ	=	360 °

Calculate fill time and Brinkman number.

Solution:
The dimensionless fill time τ which characterizes the cooling of the flowing melt in the cavity is defined as [12]

$$\tau = \frac{V \cdot a}{\dot{Q} \cdot b^2} \tag{9.2.1}$$

Inserting the numerical values we have

$$\tau = \frac{160 \cdot 7.72 \cdot 10^{-4}}{160 \cdot (0.105)^2} = 0.07$$

The Brinkman number which characterizes viscous heating is given by

$$Br = \frac{b^2 \cdot kr}{10^4 \cdot \lambda \cdot (T_M - T_W)} \cdot \left[\frac{\dot{Q} \cdot 360}{N \cdot Q \cdot 2\pi \cdot b^2 \cdot r_2}\right]^{(1+n_r)} \tag{9.2.2}$$

Introducing the given values we get

$$Br = \frac{0.105^2 \cdot 3.05 \cdot 10^4}{10^4 \cdot 0.174 \cdot 195} \left(\frac{160 \cdot 360}{1 \cdot 360 \cdot 2\pi \cdot (0.105)^2 \cdot 12}\right)^{1.2655} = 0.771$$

Table 9.7B shows the results of simulating the mold filling of a large rectangular open container (Fig. 9.11A) [24] according to the model of Stevenson [12].

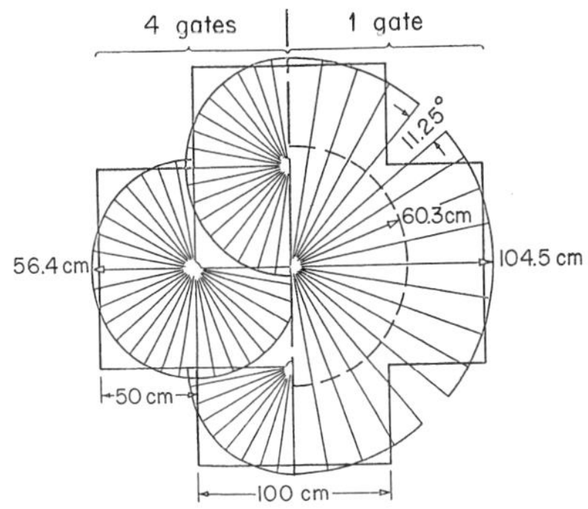

Fig. 9.11A: Mold filling of rectangular open container according to the model of Stevenson [12]

Relative to the standard conditions the results in Table 9.7B show that Δp is decreased by 10 percent when T_1 is increased 20 °C, by 4 percent when T_w is increased 20 °C, by 24 percent when the material is changed, by 37 percent when the number of gates is increased to four, and by 63 percent when all of the above changes are made. The optimal conditions which correspond to the lowest Δp and F occur when these changes are combined with the change of the resin.

Table 9.7B: Simulations of Injection Pressure and Clamp Force at Different Processing Conditions [24]

Resin	N	T_1 °K	T_w °K	Br	Δp MPa	F kN
A	1	501	301	1.381	302	191000
A	1	520	301	0.962	273	175000
A	1	500	321	1.536	290	181000
B	1	500	301	0.849	229	147000
A	4	500	301	0.625	192	67800
B	4	500	301	0.299	113	40400

Symbols in Table 9.7B:

T_1: inlet melt temperature
T_w: mold wall temperature
Δp: injection pressure drop
F: clamp force
N: Number of gates

Operating Conditions:

injection rate = 2000 cm^3/s
fill time = 3.81 s
half-thickness of the part = 0.127 cm
dimensionless fill time $\tau = 0.1824$

A rough estimate of the clamp force can be made as follows:
The pressure drop in a plate mold Δp_{IN} for the melt flow of an isothermal, Newtonian fluid can be calculated from Eqns. (7.3.2) and (7.3.4) given in Section 7.3 by putting n = 1. Δp_{IN} is then

$$\Delta p_{IN} = \frac{12 \cdot \dot{Q} \cdot \eta \cdot L}{W \cdot H^3} \tag{9.2.3}$$

Substituting the injection rate for the volume flow rate $\dot{Q}0$ we get

$$\dot{Q} = \frac{v}{t} \tag{9.2.4}$$

where $v = W \cdot H \cdot L$
and t = injection time = L/u
with u = injection velocity
Introducing the equations above into Eq. (9.2.3) we obtain

$$\Delta p_{IN} = \frac{12 \cdot L}{H} \cdot \frac{1}{H} \cdot u \cdot \eta \tag{9.2.5}$$

Example with symbols and units in Eq. (9.2.5)

part thickness	$H = 1$ mm
flow length	$L = 100$ mm
injection velocity	$u = 100$ mm/s
melt viscosity	$\eta = 250$ P · s

Solution:

$$\Delta p_{IN} = 12 \cdot \frac{100}{1} \cdot \frac{1}{1} \cdot \frac{100 \cdot 250}{10^5} = 300 \text{ bar}$$

In Fig. 9.11C Δp_{IN} is plotted over the part thickness H for different ratios of flow length L to part thickness H [25]. The injection velocity is assumed to be 10 cm/s and the melt viscosity 250 Pa · s.

The pressure Δp_{IN} is then corrected by the following correction factors [26]:

Correction Factor CF_{shape}

Depending on the complicacy of the part geometry the factor CF_{shape} can have values between 1.1 to 1.3.

Correction Factor CF_{resin}

The correction factors for the flow behavior of the resin CF_{resin} are given in Table 9.7C for some resins.

Table 9.7C: Correction Factor CF_{resin} [26]

Resin	CF_{resin}
PE, PP, PS	1.0
PMMA, PPO	1.5
PA, SB	1.2
PC, PVC	1.7
ABS, SAN, CA	1.3

Finally the actual cavity pressure Δp

$$\Delta p = \Delta p_{IN} \cdot CF_{shape} \cdot CF_{resin}$$

and the clamp force F [kN]

$$F = \frac{\Delta p(\text{bar}) \cdot \text{projected area of the part } (\text{cm}^2)}{100}$$

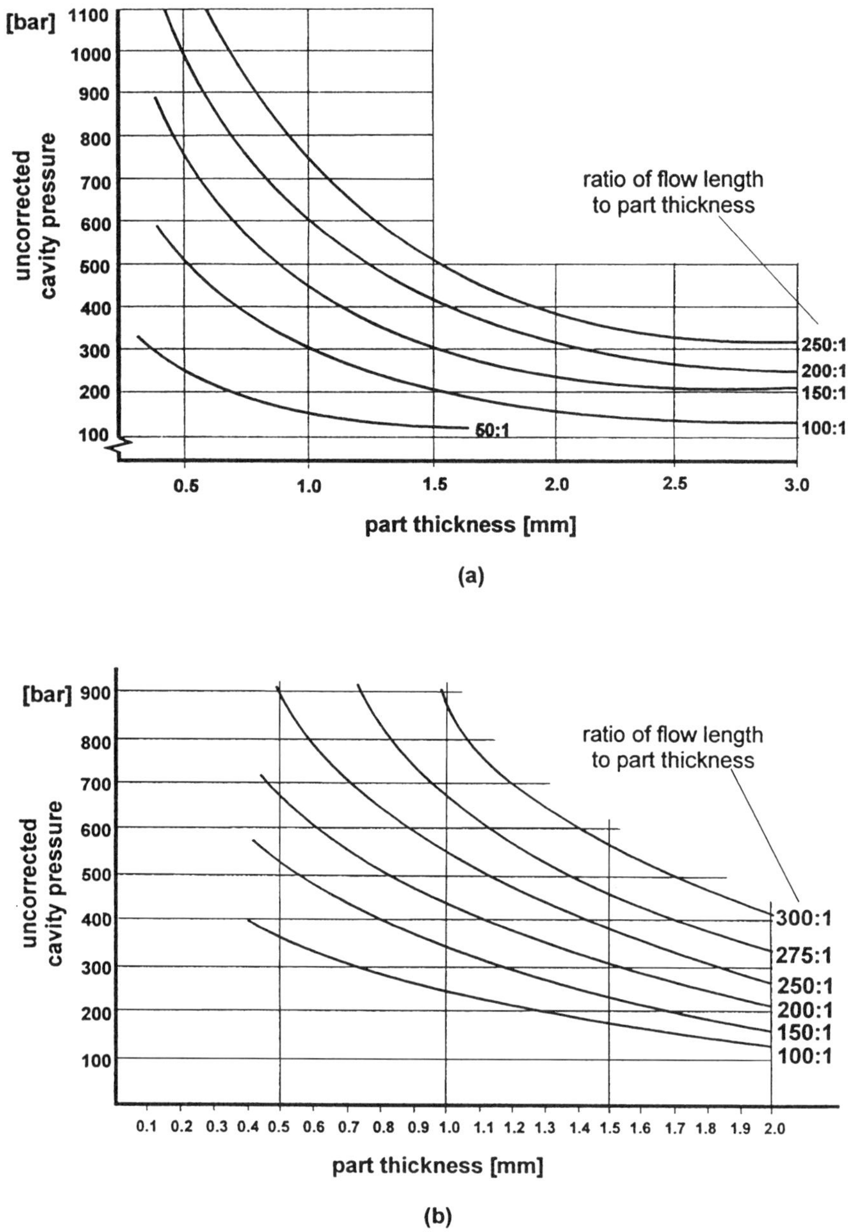

Fig. 9.11B: Cavity pressure as a function of part thickness and ratio of flow length to part thickness [25]

Example:
Disk-shaped mold (see Fig. 9.11A)

Thickness	$H = 1$ mm
Radius	$R = 100$ mm
Correction factor	$CF_{shape} = 1.2$

Resin: ABS
CF_{resin} from Table $CF_{resin} = 1.3$

Solution:
Ratio of flow length to part thickness L/H

$$L/H = 100$$

Δp_{IN} from Fig. 4.11B for $L/H = 100$ and $H = 1$

$$\Delta p_{IN} = 300 \text{ bar}$$

Actual cavity pressure Δp

$$\Delta p = 300 \cdot 1.2 \cdot 1.3 = 468 \text{ bar}$$

Clamp force F (kN)

$$F = \frac{468 \cdot 314}{100} \approx 1470 \text{ kN}$$

In Fig. 9.12 the relationship between shot size and clamping force is presented for some commercially available injection molding machines [1]. Here a and a_0 are the ratios between working (injection) capacity and clamping force and F_{cx} and F_{co} the clamping forces.

The following example shows as to how the maximum shot weight can be calculated if the clamping force is known.

Example:

Clamping force:	3500 kN
Value of the ordinate corresponding to the dashed line 3 in Fig. 9.12b:	0.39
Maximum shot weight for polystyrene assuming a density of 0.9:	3500 x 0.39 x 0.9 ≈ 1230 g
Effective maximum shot weight with a factor of 0.8:	1230 x 0.8 ≈ 980 g

The range for the maximum shot weight is approximately [1] 1230 ± 500 g.

Two-dimensional and three-dimensional diagrams as shown in Fig. 9.13 [18] are helpful in the practice for finding out the useful working range of an injection molding machine [18].

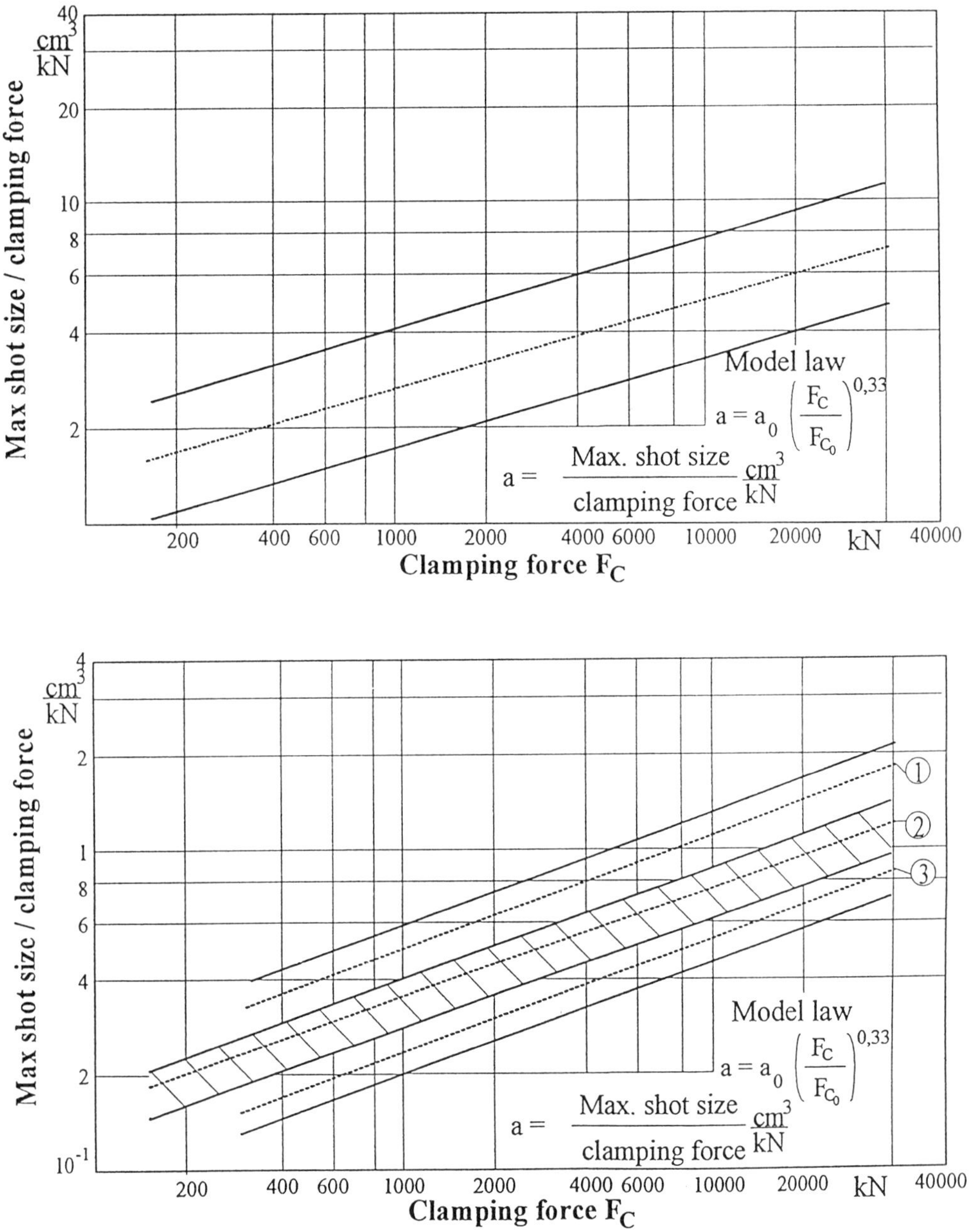

Fig. 9.12: Calculation of maximum shot weight by means of model laws [1]

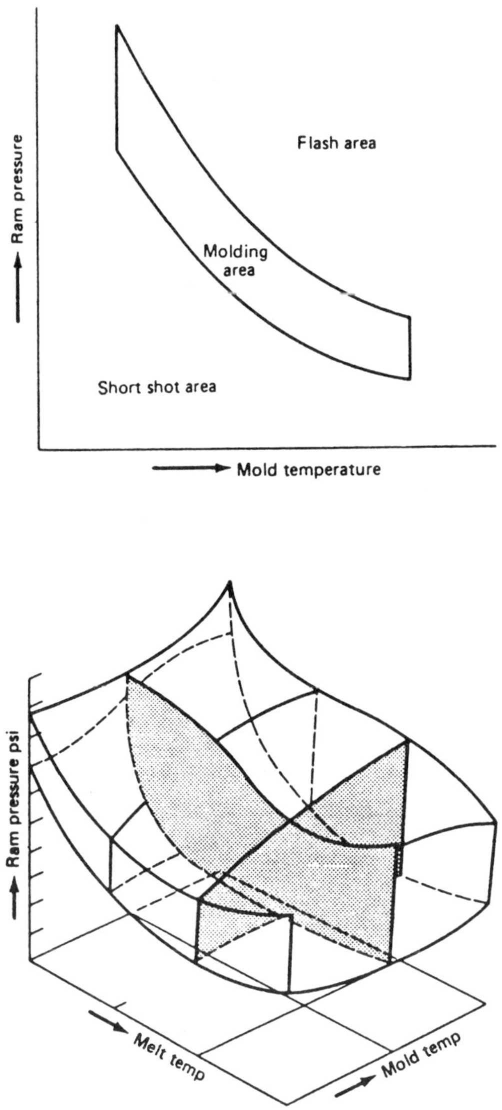

Fig. 9.13: Mold Area Diagram (MAD) and Mold Volume Diagram (MVD) [18]
a: MAD, b: MVD

9.2.3.4 Thermal Design of Mold

Cooling of Melt in Mold

The analytical solution for the transient one-dimensional heat conduction for a plate can be written as [13]

$$t_K = \frac{s^2}{\pi^2 \cdot a} \cdot \ln\left[\frac{4}{\pi}\left(\frac{T_w - T_a}{T_w - T_b}\right)\right] \tag{9.2.6}$$

Although this equation is valid only for amorphous polymers, it has given good results in the practice for semi-crystalline and crystalline polymers as well, so that its general use for all thermoplastic materials is justified.

Example with symbols and units:
Cooling of a part in an injection mold for the following conditions [5]:
resin: PE-LD

thickness of the plate	s	=	12.7 mm
temperature of the melt	T_a	=	243.3 °C
mold temperature	T_w	=	21.1 °C
demolding temperature	T_b	=	76.7 °C
thermal diffusivity of the melt	a	=	$1.29 \cdot 10^{-3}$ cm²/s

The temperature T_b in Eq. (9.2.6) refers to the centerline temperature of the part.

Solution:
The temperature ratio θ_{T_b}:

$$\theta_{T_b} = \frac{T_b - T_w}{T_a - T_w} = \frac{76.7 - 21.1}{243.3 - 21.1} = 0.25$$

cooling time t_k according to Eq. (9.2.6)

$$t_k = \frac{1.27^2}{\pi^2 \cdot 1.29 \cdot 10^{-3}} \cdot \ln\left(\frac{4}{\pi} \cdot \frac{1}{0.25}\right) = 206.5 \text{ s}$$

The cooling time t_k can also be found from Fig. 9.14 in which the ratio θ_{T_b} is plotted over the Fourier number for bodies of different geometry [14].
Fourier number F_o from Fig. 9.14 at $\theta_{T_b} = 0.25$

$$F_o = 0.66$$

Cooling time t_k:

$$x = \frac{s}{2} = \frac{12.7}{2} = 6.35 \text{ mm}$$

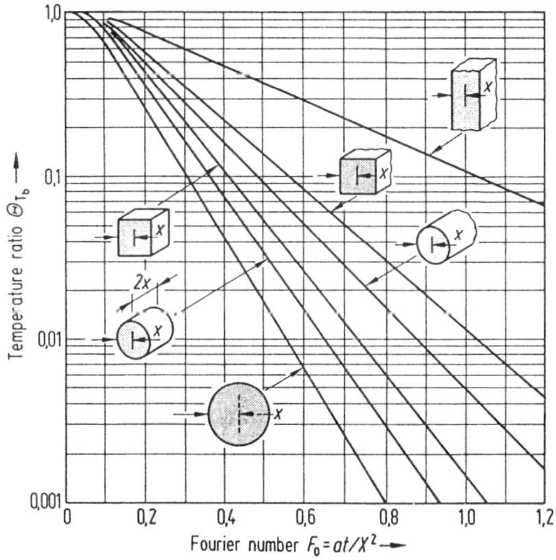

Fig. 9.14: Axis temperature for multidimensional bodies [14]

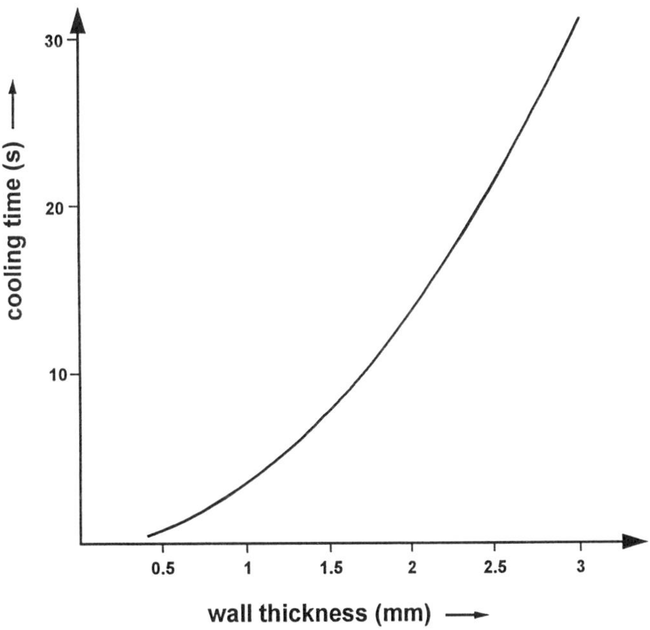

Fig. 9.15: Cooling time vs. wall thickness

$$F_o = \frac{a \cdot t_k}{x^2} = \frac{1.29 \cdot 10^{-3} \cdot t_k}{(0.635)^2} = 0.66$$

$$t_k \approx 206 \text{ s}$$

The cooling time as a function of wall thickness of the molding is shown in Fig. 9.15.

Calculations of cooling time based on the average demolding temperature are given in the book by Rao [4].

Convective heat transfer is to be taken into account if the mold is cooled by a coolant such as circulating water by calculating the Biot number as shown in Chapter 7.

In practice the temperature of the mold wall is not constant, as it is influenced by the heat transfer between the melt and the cooling water. Therefore the geometry of the cooling channel lay-out, thermal conductivity of the material and velocity of cooling water affect the cooling time significantly (Fig. 9.16).

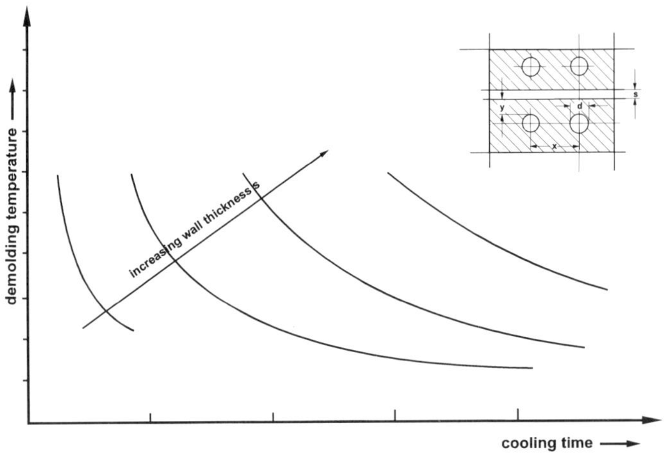

Fig. 9.16: Cooling time as a function of demolding temperature and wall thickness

The influence of the cooling channel lay-out on cooling time can be simulated by means of equations given in the book [4] by changing the distances x and y (Fig. 9.16) as shown in Figs. 9.17 and 9.18. The effects of the temperature of cooling water and of water velocity are presented in Figs. 9.19 and 9.20 respectively. From these results it follows that the cooling time is significantly determined by the cooling channel lay-out.

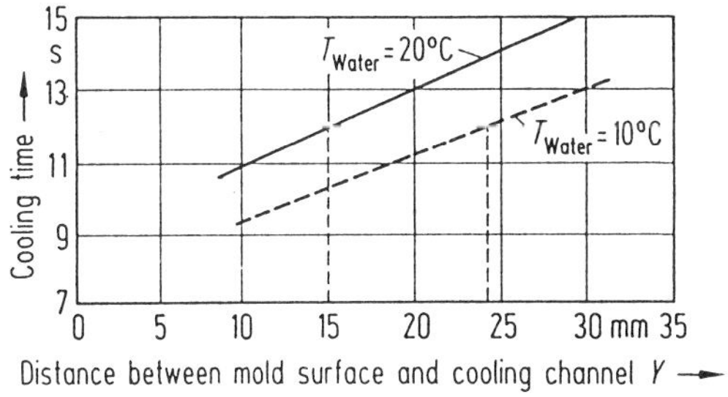

Fig. 9.17: Effect of cooling channel distance Y on cooling time [4]

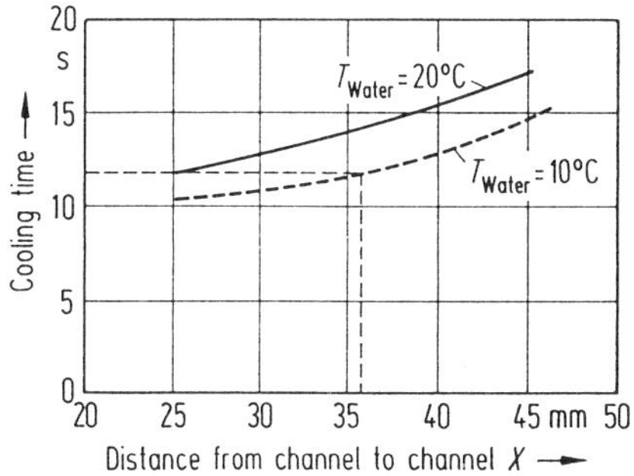

Fig. 9.18: Effect of cooling channel distance X on cooling time [4]

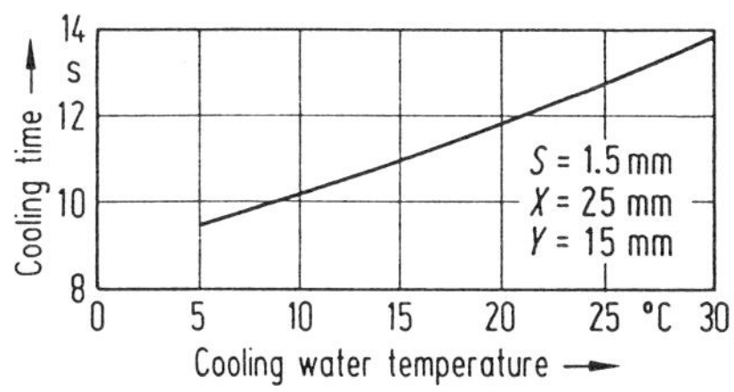

Fig. 9.19: Influence of the temperature of cooling water on cooling time [4]

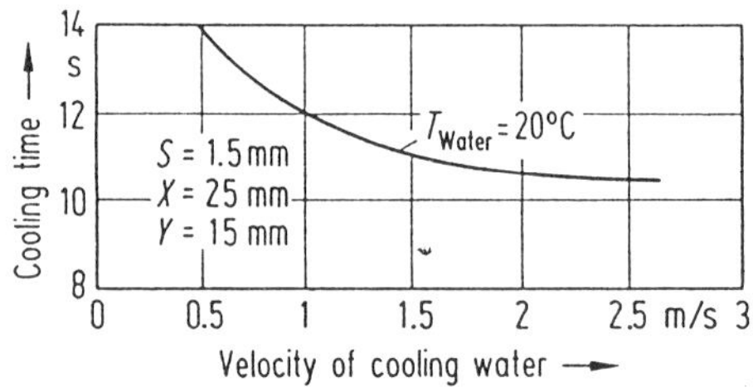

Fig. 9.20: Influence of the velocity of cooling water on cooling time[4]

9.2.3.5 Mechanical Design of Mold

The cooling channels should be as close to the surface of the mold as possible so that heat can flow out of the melt in the shortest time possible. However, the strength of the mold material sets a limit to the distance between the cooling channel and the mold surface. The allowable distance z (Fig. 9.21) taking the strength of the mold material into account was calculated by Lindner [15] according to the following equations:

$$\sigma_b = \frac{0.5 \cdot p \cdot d^2}{z^2} \tag{9.2.7}$$

$$\tau = \frac{0.75 \cdot p \cdot d}{z} \tag{9.2.8}$$

$$f = \frac{1000 \cdot p \cdot d^2}{z} \left(\frac{d^2}{32 \cdot E \cdot z^2} + \frac{0.15}{G} \right) \tag{9.2.9}$$

where p = melt pressure N/mm^2
d, z = distances mm, see Fig. 9.21
E = tensile modulus N/mm^2
G = shear modulus N/mm^2
σ_b = tensile stress N/mm^2
τ = shear stress N/mm^2
f = deflection of the mold material above the cooling channel μm

The minimization of the distance z such that the conditions

$$f \leq f_{max}$$

$$\sigma_b \leq \sigma_{bmax}$$

$$\tau \leq \tau_{max}$$

(f_{max}: maximum deflection; σ_{bmax}: allowable tensile stress; τ_{max}: allowable shear stress) are satisfied, can be accomplished by a computer program [16]. The results of a sample calculation are shown in Table 9.8.

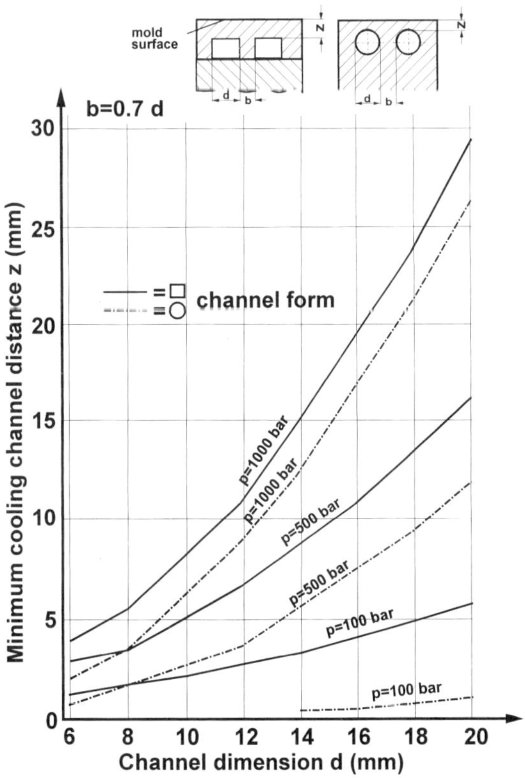

Fig. 9.21: Cooling channel distance z for different cavity pressures and channel dimensions with steel as mold material

Table 9.8: Results of Optimization of Cooling Channel Distance

Input	Output
melt pressure p = 4.9 N/mm^2 maximum deflection f_{max} = 2.5 μm modulus of elasticity E = 70588N/mm^2 modulus of shear G = 27147 N/mm^2 allowable tensile stress σ_{bmax} = 421.56 N/mm^2 allowable shear stress τ_{max} = 294.1 N/mm^2 channel dimension d = 10 mm	channel distance z = 2.492 mm deflection f = 2.487 μm tensile stress σ = 39.44 N/mm^2 shear stress τ = 14.75 N/mm^2

The equations given above are approximately valid for circular channels as well. The distance from wall to wall of the channel in this case should be roughly around channel length l or channel diameter depending on the strength of mold material.

Fig. 9.21 shows for steel as mold material the minimum cooling channel distance z for different cavity pressures.

Literature

1. *Johannaber, F.:* Injection Molding Machines, Hanser, Munich 1983
2. *McKelvey, J.M.:* Polymer Processing, John Wiley, New York 1962
3. *Hensen, F. (Ed.):* Plastics Extrusion Technology, Hanser, Munich 1988
4. *Rao, N.S.:* Design Formulars for Plastics Engineers, Hanser, Munich 1991
5. *Thorne, J.L.:* Plastics Process Engineering, Marcel Dekker Inc., New York 1979
6. *Rao, N.S.:* Computer Aided Design of Plasticating Screws, Hanser, Munich 1986
7. *Donovan, R.C.:* Polymer Engineering and Science 11 (1971) p. 361
8. *Rao, N.S., O'Brien, K.T. and Harry, D.H.:* Computer Modeling for Extrusion and other Continuous Polymer Processes. Ed. Keith T. O'Brien, Hanser, Munich 1992
9. *Austin, C.:* CAE for Injection Molding, Hanser, Munich 1983
10. *CADMOULD:* IKV Aachen
11. *Bangert, H.:* Systematische Konstruktion von Spritzgießwerkzeugen unter Rechnereinsatz, Dissertation RWTH Aachen 1981
12. *Stevenson, J.F.:* Polymer Engineering and Science 18 (1978), 9. 573
13. *Martin, H.:* in VDI-Wärmeatlas, VDI-Verlag, Düsseldorf 1984
14. *Welty, J.R., Wicks, C.E., Wilson, R.E.:* Fundamentals of Momentum, Heat and Mass Transfer, John Wiley, New York 1983
15. *Lindner, E.:* Berechenbarkeit von Spritzgießwerkzeugen, VDI-Verlag, Düsseldorf 1974, p. 109
16. *Rao, N.S.:* Designing Machines and Dies for Polymer Processing with Computer Programs, Hanser, Munich 1981
17. *Zoellner, O., Sagenschneider, U.:* Kunststoffe 84 (1994) 8, p. 1022
18. *Rosato, D.V., Rosato, D.V.:* Plastics Processing Data Handbook, Van Nostrand-Reinhold, New York 1990
19. *C-Mold:* Advanced CAE Technology Inc.
20. *Menges, G., Mohren, P.:* How to make Injection Molds, Hanser, Munich 1993
21. *Brochure of Spirex Corporation* U.S.A., 1995
22. *Pötsch, G., Michaeli, W.:* Injection Molding - An Introduction, Hanser, Munich 1995
23. *Cracknell, P.S., Dyson, R.W.:* Handbook of Thermoplastics Injection Mold Design, Blackie Academic & Professional, London 1993
24. *Stevenson, J.F., Hauptfleisch, R.A.:* Handbook Methods for Predicting Pressure Drop and Clamp Force in Injection Molding: Technical Report No. 7, July 30, 1976
25. *Brochure of Mannesmann-Demag* 1992

A FINAL WORD

There are many other processes in wide application such as reaction injection molding, compression and transfer molding used to produce parts from thermosetting resins. It was not possible to cover all these processes within the scope of this book. However, the nature of data needed for designing machinery for these processes is similar to the one presented in this work.

Index

Authors

Natti S. Rao obtained his B.Tech(Hons) in Mechanical Engineering and M.Tech in Chemical Engineering from the Indian Institute of Technology, Kharagpur, India. After receiving his Ph.D. from the University of Karlsruhe,Germany he worked for the BASF AG for a number of years. Later on he served as a technical advisor to the leading German Plastics machine manufacturers.
Natti has published over 50 papers on different aspects of plastics engeneering and authored 3 books on designing polymer machinery with computers. Prior to starting his consultancy company in 1987, he worked as a visiting professor at the Indian Institute of Technology, Chennai(Madras), India. Besides consultancy work, he also holds seminars teaching the application of his software for designing extrusion and injection molding machinery.

Keith O´Brien has a B.Sc.(Eng.) from the University of London, and M.Sc. and Ph.D. degrees from the University of Leeds in Mechanical Engineering. Keith has over 100 published works to his credit, and also edited the well-known book, ' Computer Modeling for Extrusion and other Continuous Processes'. Before he joined Johnson & Johnson , he was professor of mechanical engineering at the New Jersey Institute of Technology.